SO-AEE-419

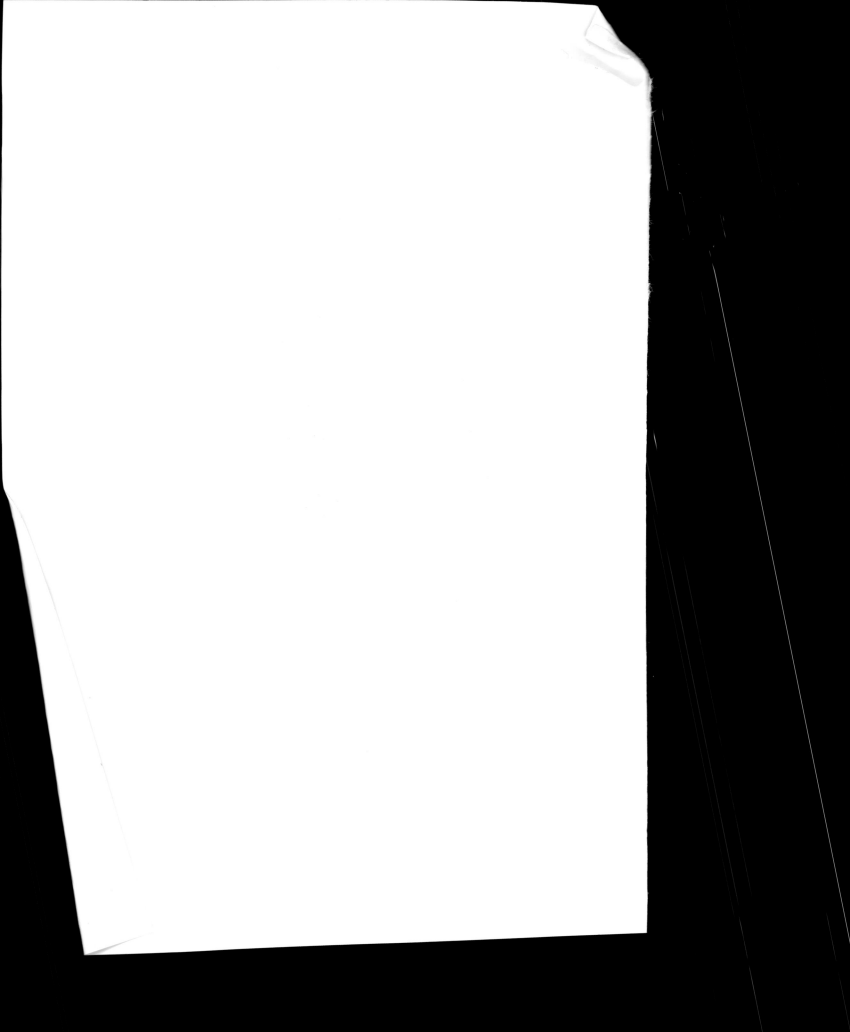

MOLECULAR
BIOLOGY
INTELLIGENCE
UNIT

CELL-MEDIATED EFFECTS OF IMMUNOGLOBULINS

Wolf Herman Fridman, M.D., Ph.D.
Catherine Sautès, Ph.D.

Institut Curie
Paris, France

CHAPMAN & HALL
ITP An International Thomson Publishing Company

New York • Albany • Bonn • Boston • Cincinnati • Detroit • London • Madrid • Melbourne •
Mexico City • Pacific Grove • Paris • San Francisco • Singapore • Tokyo • Toronto • Washington

AUSTIN, TEXAS

MOLECULAR BIOLOGY INTELLIGENCE UNIT
CELL-MEDIATED EFFECTS OF IMMUNOGLOBULINS

R.G. LANDES COMPANY
Austin, Texas, U.S.A.

Please address all inquiries to the Publishers:
R.G. Landes Company, 810 S. Church Street, Georgetown, Texas, U.S.A. 78626
Phone: 512/ 863 7762; FAX: 512/ 863 0081

North American distributor:
Chapman & Hall, 115 Fifth Avenue, New York, New York, U.S.A. 10003

CHAPMAN & HALL

U.S. and Canada ISBN: 0-412-11821-1

Library of Congress Cataloging-in-Publication Data

Cell-mediated effects of immunoglobulins / [edited by] Wolf H. Fridman, C. Sautès.
        p. cm. -- (Molecular biology intelligence unit)
    Includes bibliographical references and index.
    ISBN 1-57059-396-5 (alk. paper)
    1. Immunoglobulins. 2. Fc receptors. I. Fridman, Wolf H. (Wolf Herman) II. Sautès, Catherine. III. Series.
    [DNLM: 1. Immunoglobulins. 2. Immunity, Cellular. QW 601 C3927 1996]
QR186.7.C43 1996
616.07'93--dc20
DNLM/DLC
for Library of Congress

96-33465
CIP

PUBLISHER'S NOTE

R.G. Landes Company publishes six book series: *Medical Intelligence Unit, Molecular Biology Intelligence Unit, Neuroscience Intelligence Unit, Tissue Engineering Intelligence Unit, Biotechnology Intelligence Unit* and *Environmental Intelligence Unit.* The authors of our books are acknowledged leaders in their fields and the topics are unique. Almost without exception, no other similar books exist on these topics.

Our goal is to publish books in important and rapidly changing areas of bioscience and environment for sophisticated researchers and clinicians. To achieve this goal, we have accelerated our publishing program to conform to the fast pace in which information grows in bioscience. Most of our books are published within 90 to 120 days of receipt of the manuscript. We would like to thank our readers for their continuing interest and welcome any comments or suggestions they may have for future books.

Shyamali Ghosh
Publications Director
R.G. Landes Company

CONTENTS

EDITORS

Wolf Herman Fridman
Laboratoire d'Immunologie Cellulaire et Clinique
INSERM U255
Institut Curie
Paris, France
Chapter 1

Catherine Sautès
Laboratoire d'Immunologie Cellulaire et Clinique
INSERM U255
Institut Curie
Paris, France
Chapters 2, 6

CONTRIBUTORS

Sebastian Amigorena
Biologie Cellulaire de la
  Présentation Antigénique,
  INSERM CJF95-01
Paris, France
Chapter 5

Christian Bonnerot
Biologie Cellulaire de la
  Présentation Antigénique,
  INSERM CJF95-01
Institut Curie
Paris, France
Chapter 3

Marc Daëron
Laboratoire d'Immunologie
  Cellulaire et Clinique,
  INSERM U255
Institut Curie
Paris, France
Chapter 4

Jean-Luc Teillaud
Laboratoire d'Immunologie
  Cellulaire et Clinique,
  INSERM U255
Institut Curie
Paris, France
Chapter 7

CHAPTER 1

STRUCTURE AND FUNCTION OF IMMUNOGLOBULINS

Wolf Herman Fridman

Over a century ago (1890), Emil Behring and Shibasaru Kitasato observed that serum taken from rabbits injected with bacterial toxins could neutralize these toxins.[1] Then it was shown that passive transfer of serum from a rabbit immunized with diphtheria toxin could protect a naive animal from diptheria infection. Several authors demonstrated that the protecting activity of immune serum could be attributed to a particular group of serum proteins that were called antibodies.[1]

During the 20th century, the development of biochemical technology resulted in an explosion of knowledge on the molecular structures of antibodies, at the primary, secondary and tertiary levels. Since they mediate immune reactions, antibodies received the generic name of immunoglobulins (Ig).

Over the last 30 years, the structural bases of immunoglobulin functions were established, the cells that produce them (the B lymphocytes) were identified, and the molecules and cells with which they interact were characterized.

STRUCTURAL BASES OF ANTIBODY FUNCTIONS

All immunoglobulins, regardless of their class or their antibody specificity, have a prototypic structure, composed of four polypeptide chains (reviewed in ref. 2). Two identical heavy (H) chains with a molecular mass of 50-60 kDa are linked between them by disulfide bridges. Each H chain is linked to a light (L) chain, of 25 kDa, also by a disulfide bridge (Fig. 1.1). Polymeric immunoglobulins are formed by the association of several (2 to 5) prototype units, with additional peptide chains to hold them together.

Cell-Mediated Effects of Immunoglobulins, edited by Wolf Herman Fridman and Catherine Sautès. © 1997 R.G. Landes Company.

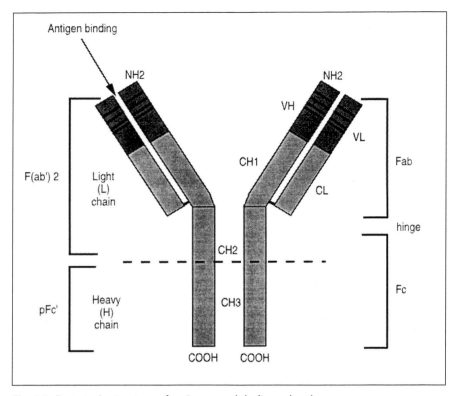

Fig. 1.1. Prototypic structure of an immunoglobulin molecule.

Immunoglobulins are glycoproteins with several carbohydrate chains, usually bound to their H chains. In addition to immunoglobulins that circulate in biological fluids, a subset of lymphocytes also bear immunoglobulin at their surface. These are the B lymphocytes, which upon maturation, activation and differentiation, become the antibody-producing cells. On these cells, immunoglobulins are receptors for antigen and contain an additional peptide stretch at the COOH-terminal end of their H chains, that anchor them in the cell membrane.[3,4] Immunoglobulins are a hallmark of the immune system of all mammals.

BASIC IMMUNOGLOBULIN STRUCTURE

The basic four-chain model of immunoglobulin molecules is schematized in Figure 1.1. The COOH-terminal half of the light chain is constant, except for allotypic (discriminating between individuals within a species) and isotypic (see 1-2) variations. It is therefore called CL, for constant light chain. The NH_2-half of the L chain exhibits high sequence variability within the antibodies of an individual, and is therefore called VL, for variable light chain. CL and VL are encoded by different C and V genes. The heavy chain is constructed following the same model as the light chain. A large COOH-terminal part, representing three quarters of the molecule, is constant, except for allotypic and isotypic variations. It is called CH, for constant heavy chain. The NH_2-terminal part of the chain contains high sequence variability and is therefore called VH, for variable heavy chain. As for light chains, CH and VH are encoded by different C and V genes.[5] Both in heavy and light chains, the NH_2-terminal domain corresponds to the variable region, VH and VL respectively. The light chain has one constant domain (CL) while the heavy chain is composed of three (CH1, CH2 and CH3) to four (CH4) constant domains. X-ray crystallography have shown that Ig molecules have a Y-shaped structure (Fig. 1.2) that has also been visualized by electron microscopy.

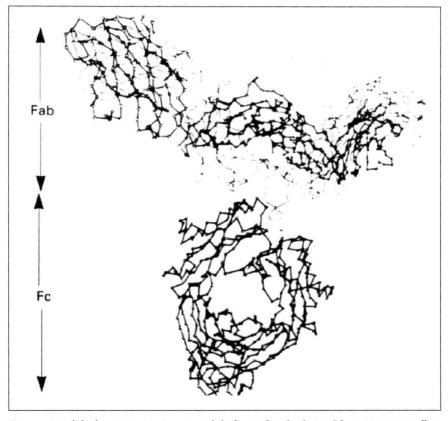

Fig. 1.2. Model of a prototype immunoglobulin molecule derived from X-ray crystallography. Reprinted with permission from: Roitt I, Brostoff I, Male D. Mosby, 3rd edition, 1990.

Analysis of the secondary and tertiary structures of immunoglobulins revealed that they are composed of globular regions of similar shapes. These regions define domains that are formed by intra-chain disulfide bounds. Each domain contains about 110 amino acids and is characteristic, not only of immunoglobulin, but also of the members of a vast genetic family, known as the immunoglobulin gene superfamily.[6]

IMMUNOGLOBULIN FUNCTIONAL SITES

The treatment of an immunoglobulin molecule with papain, a proteolytic enzyme, produces two fragments: an antigen-binding component, called Fab (for antigen-binding) and a crystallizable fragment devoid of antigen binding capacity called Fc, (for crystallizable). A Fab fragment is composed of the association of a light chain with the VH and CH1 domains of an heavy chain. It contains the antigen-binding site formed by non-covalent association of one VL and VH region. The association of a given variable region of a light chain with a given variable region of a heavy chain defines the antigen-specificity of an antibody. An Fc fragment is composed of the COOH-terminal domains of the heavy chains (CH2 and CH3, sometimes CH4) which are linked together by interchain disulfide bridges in the hinge region which lies between the Fab and Fc fragments (Fig. 1.1). The hinge region gives flexibility to the immunoglobulin molecule allowing it to reach the best conformation to bind both to antigen with high affinity as well as to effector molecules and cells.[7,8]

The treatment of an immunoglobulin molecule with pepsin, another proteolytic enzyme which cleaves at sites COOH-terminal of the hinge region, generates a F(ab')2 fragment composed of two Fab fragments linked by the disulfide bridges of the hinge region and a short pFc' fragment corresponding to the COOH-part of an Fc region. In contrast to a Fab fragment that has one antigen-binding site, F(ab')2 fragments have two antigen-binding sites as a complete prototype immunoglobulin. They bind to antigen with an affinity similar to that of whole immunoglobulin but lack functional activities, which indicates that the Fc fragment is required for antibodies to be biologically active.

Immunoglobulins are bifunctional molecules. One region binds foreign, or self, antigens and ensures the specificity of the antibody reaction while the other, the Fc part, mediates effector functions upon interaction with host components such as cells of the immune system, phagocytes, or components of a serum enzymatic system, called complement. Within an individual, there are populations of immunoglobulins with distinct properties to interact with complement or host cells, which are therefore endowed with distinct biological functions. These populations have been molecularly and functionally characterized and are referred to as Ig classes and sub-classes or immunoglobulin isotypes.

IMMUNOGLOBULIN ISOTYPES

Isotypes are defined by the existence of distinct genes which code for a constant region of an immunoglobulin. In mammals, there are two light chain isotypes, called κ an λ and up to nine heavy chain isotypes, called in human μ, λ, γ1, γ2, γ3, γ4, α1, α2 and ε. Each immunoglobulin is composed of L chains of one or the other type (κ or λ) and H chains of a given isotype, which defines the immunoglobulin class. In mammals, 5 classes have been identified: IgM with heavy chains encoded by the Cμ gene, IgD encoded by the Cδ gene, IgG encoded by Cγ genes, IgA encoded by Cα genes and IgE encoded by the Cε gene. Since there are up to four Cγ genes—in man and mouse— and two Cα genes in humans, they define IgG and IgA sub-classes, respectively. Within a class, sub-class structures are more similar than between classes.

Classes and sub-classes of immunoglobulins are often referred to as isotypes, defined by the isotype of their heavy chain, regardless of the isotype of their light chain. In humans, there are nine immunoglobulin isotypes: IgM, IgD, IgG1, IgG2, IgG3, IgG4, IgA1, IgA2 and IgE. Mouse immunoglobulins which have often been utilized to study binding to host cells and antibody functions fall into eight isotypes: IgM, IgD, IgG1, IgG2a, IgG2b, IgG3, IgA and IgE.

IMMUNOGLOBULIN M (IGM)

IgM are high molecular weight glycoproteins (950-1150 kDa), composed of five identical subunits (Fig. 1.3), called IgM monomers. Each subunit is formed by the association of two light chains (of κ or λ isotype) to two heavy chains (of the μ isotype) in a symmetrical way. Therefore, each pentameric IgM molecule contains ten heavy chains, ten light chains, each subunit being linked to the others by a joining peptide, the J chain (Fig. 1.3). Since each subunit has two antigen-binding sites on each Fab fragment, IgM antibodies have ten putative antigen-combining sites, which confer high avidity, even if the affinity of each individual binding site is moderate. IgM also have five Fc regions that make them very powerful agents in triggering effector enzymes of the complement system.

IgM appeared early in evolution, and primitive vertebrates have macroglobulins whose structure resembles that of mammalian IgM and which may be involved in the defense of these vertebrates against infections. The IgM molecules are the first immunoglobulins synthesized by neonates and circulate in serum at a concentration of 0.7-1.7 g/L, representing 5-10% of all immunoglobulins. IgM are the preponderant immunoglobulins to be produced during the early phase of an immune response at a moderate synthesis rate of 7 mg/kg/day. They have a short half-life of 5 days (Table 1.1). Natural antibodies, present in individuals prior to the encounter of antigen, are of the IgM isotype. Finally, in a monomeric form, membrane bound IgM functions as an

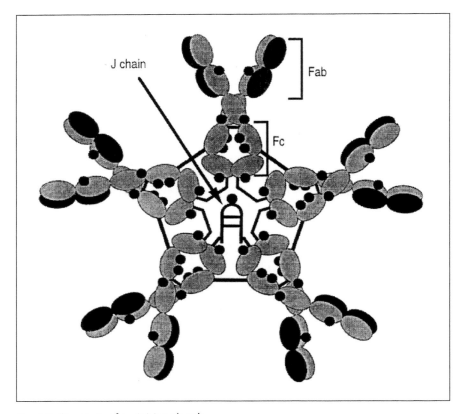

Fig. 1.3. Structure of an IgM molecule.

antigen receptor at the surface of most B cells, especially naive
B lymphocytes.

IgM antibodies are involved in the first line of immunological defense
against micro-organisms, especially bacteria. They hardly bind to cells
and do not cross epithelia, such as gut or placenta, but activate complement very efficiently, inducing a cascade of enzymes which will, eventually, destroy the bacterial wall.

IMMUNOGLOBULIN D (IgD)

IgD are monomeric molecules composed of two light chains (λ or
κ) and two heavy δ chains, organized in a symmetrical way. IgD are
hardly detectable in serum, where they represent less than 0.3% of all
immunoglobulins and have a very short half-life of three days, due to
proteolytic degradation (Table 1.1). In fact, IgD, like IgM monomers,
are essentially produced as a membrane bound form which functions
as an antigen receptor on mature, unstimulated B lymphocytes. No
biological function has been attributed to serum IgD.

Table 1.2. Physicochemical properties of human immunoglobulin isotypes

Property	IgG	IgA	IgM	IgD	IgE
Usual molecular form	Monomer	Monomer, dimer	Pentamer, hexamer	Monomer	Monomer
Other chains	None	J-chain, SC	J-chain	None	None
Sub-classes	G1, G2, G3, G4	A1, A2	None	None	None
Molecular weight	150 kDa	160 kDa, 400 kDa	950 kDa, 1150 kDa	175 kDa	190 kDa
Sedimentation constant (Sw20)	6.6S	7S, 11S	19S	7S	8S
Carbohydrate content (%)	3	7	10	9	13
Serum level (mg/ml)	9.5–12.5	1.5–2.6	0.7–1.7	<0.04	0.0003
Percentage of total serum Ig	75–85	7–15	5–10	0.3	0.019
Serum half-life (days)	23	6	5	3	2.5
Synthesis rate (mg/kg/day)	33	65	7	0.4	0.016
Antibody valence	2	2, 4	10, 12	2	2
Complement activation	+	–	++	–	–
Biological properties	Placental transfer secondary Ab for most anti-pathogen responses	secretory immunoglobulin	Primary Ab responses to microorganisms	marker for mature B cells	Allergy, anti-parasite responses

IMMUNOGLOBULIN G

IgG is the major class of immunoglobulins, accounting for about three quarters of all serum immunoglobulins. IgG molecules are monomers of 150 kDa, composed of two light chains (λ or κ) and two heavy γ chains. Their carbohydrate content is usually lower (3%) than that of other immunoglobulin isotypes (8-12%). The serum concentration of IgG is of 9.5-12.5 g/L, representing 15% of all serum proteins.

IgG antibodies are produced at very low levels in a primary immune response, but are the prominent immunoglobulins synthesized by B cells in secondary responses, upon iterative exposure to antigen. IgG synthesis rate is high, 33 mg/kg/day, and its half-life in serum is long, 33 days. Since IgG crosses the epithelial barriers, especially placenta, it makes it the major class of antibodies involved in the host defense against infection, both in individuals with a mature immune system and in neonates who are protected by the maternal IgG antibodies.

The biological functions of IgG differ partly according to sub-classes and derive from the structural differences between IgG isotypes. In man, four sub-classes have been characterized that differ in primary, secondary and tertiary structures. For the latter, the most striking difference is in the very long hinge region of IgG3, as compared to any other isotype (Fig. 1.4). The four IgG sub-classes also have different numbers and localization of their disulfide bonds (Fig. 1.4) which impose constraints on the IgG molecules resulting in different capacities to interact with effector molecules such as complement or with specific receptors expressed on various host cells. An example of such

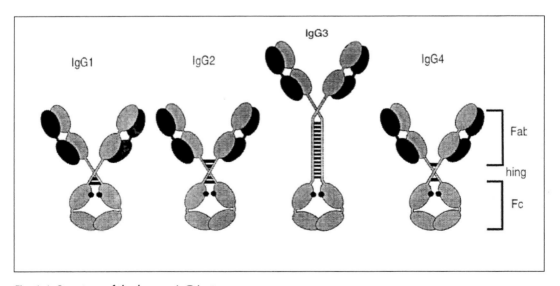

Fig. 1.4. Structure of the human IgG isotypes.

constraints is illustrated by the human IgG4 molecule. While entire IgG4 molecules are unable to activate the complement system, even when they are bound to multivalent antigens, and therefore aggregated, the removal by papain cleavage, of the Fab arms, generates an Fc fragment capable, when aggregated, of binding the first component of complement and subsequently activating the complement system. Therefore, it is not the primary sequence that binds complement which is lacking in IgG4 antibodies but the obstruction of the Fab arms that makes it unavailable.

IgG sub-classes differ slightly in their molecular mass and carbohydrate contents, but vary greatly in their serum concentration, average half-life and biological functions.

IgG1 is the predominant immunoglobulin isotype in human serum. It circulates at a concentration of c.a. 9g/L and represents over 70% of serum IgG and 50% of serum immunoglobulins. IgG1 half-life is long (20-30 days). After binding to antigen, IgG1 efficiently activates complement and binds to receptors for IgG, called Fcγ receptors (FcγR), to recruit and activate phagocytic or effector cells. IgG1 crosses the placental barrier and the major protective maternal antibody for the neonate.

IgG2 is the second most important immunoglobulin found in human serum. It circulates at a concentration of 3 g/L, representing 20% of all IgG and has a long half-life of 20-30 days. IgG2 is a poor activator of the complement system and binds with low affinity to cell-associated Fcγ receptors. As the other IgG, IgG2 crosses the placental barrier. The role of IgG2 in immune defenses is not clearly established.

IgG3 circulate in human serum at a concentration of c.a. 1g/L and have the shortest half-life among the IgG isotypes (7 days). Due to its long hinge region, IgG3 have the highest molecular mass (170 kDa as compared to 150 kDa for the other IgG isotypes). Probably also due to their elongated hinge region, IgG3 antibodies are most effective in triggering effector functions. After binding to antigen, IgG3 activate complement with a potency similar to antigen and bind to Fcγ receptors on cells with strong affinity. IgG3 also crosses the placental barrier and plays a major role in neonate protection against infection.

IgG4 is a rare immunoglobulin found at a concentration less than 0.5 g/L in human serum, but which exhibits the longest half-life of all immunoglobulins (25-35 days). IgG4 structure is characterized by the fact that it has no hinge region between the Fab and Fc parts of the molecules, which results with a loss of binding to the first component of complement and to cell-associated Fcγ receptors IgG4 crosses the placental barrier. Its role is not elucidated but its increased levels in some allergic states suggest that it may exert, yet undefined, activities in inflammatory processes.

IMMUNOGLOBULIN A (IGA)

IgA is the second major class of immunoglobulins. IgA molecules are found either as monomers or as polymers, mainly dimers, but sometimes trimers and tetramers. Each IgA monomer is composed of two light chains (λ or κ) associated with two heavy α chains. Each heavy chain is arranged in four domains (VH, CH1, CH2, CH3) along with a COOH-terminal 18 amino acids peptide; the penultimate cysteine covalently binds a joining J chain to form dimers (Fig. 1.5).

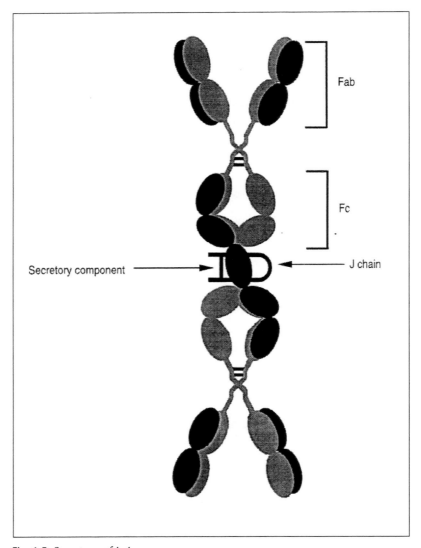

Fig. 1.5. Structure of IgA.

IgA are found in serum at a concentration of 1.5-2.6 g/L, where they represent 15% of all immunoglobulins, with a short half-life of 6 days. However IgA is the predominant immunoglobulin to be produced in the body, at a rate of 65 mg/kg/day, mostly in the gut and exocrine secretions such as saliva, colostrum, breast, milk and tears. Secretory IgA differ from serum monomeric IgA. Secretory IgAis usually composed of two IgA monomers linked by a J chain and associated with a secretory component forming a molecule of 380 kDa (Fig. 1.5). In contrast to the 15 kDa J chain which is synthesized by the antibody producing cells together with IgA, the 70 kDa secretory component is produced by epithelial cells. The manner in which a secretory IgA molecule is formed is seen as follows:

In the submucosal layer, plasma cells synthesize both IgA and J chains which allow the production of dimeric IgA molecules. The latter actively bind the secretory component when they cross the epithelial cell layers. The secretory component, then, facilitates the transport of IgA into secretions and protects it from proteolytic digestion, prolonging its half-life in secretions.

In man, two IgA sub-classes have been characterized. IgA1 is predominantly found in serum whereas IgA2 predominates in secretions. There is no evident difference in the biological activities of IgA1 and IgA2, but the former isotype is highly susceptible to proteolytic enzymes released by microorganisms in the respiratory and gastrointestinal tracts, which may explain why IgA1 is not found in secretions. IgA do not activate complement but bind with high affinity to specific receptors on cells of the myeloid and monocytic lineage, such as neutrophils and macrophages, inducing phagocytosis of microorganisms to which they are bound and triggering the production of bactericidal substances by neutrophils and macrophages.

IgA is the immunoglobulin isotype which protects the body against mucosal infections. It does not cross the placental barrier.

IMMUNOGLOBULIN E

Under normal conditions, only traces of IgE are found in human serum. IgE are monomeric immunoglobulins composed of two light chains (κ or λ) and two heavy ε chains arranged in a symmetrical way. The IgE heavy chain is arranged in five domains (VH, CH1, CH2, CH3, CH4) resulting in a molecular mass of 190 kDa (Fig. 1.6).

Two types of chronic stimulation of the immune system induce a tremendous (several hundred times) increase in IgE production. The first set of stimuli are known as allergens since they induce allergic reactions such as hay fever or asthma. Not only is the production of IgE the hallmark of allergy but, despite its mediocre serum life (2-5 days), it is the mediator of allergic reactions. IgE does not cross the placental barrier, nor does it activates complement, but it binds as monomers, in the absence of antigen, to high affinity receptors expressed mainly on mast cells and basophils. When an allergen penetrates the body, it binds to these cell-associated IgE which become aggregated and trigger the release of inflammatory mediators, such as histamine, serotonin or leukotrienes.[9]

The second type of stimulus is delivered during several parasitic infections, such as *Schistosoma mansoni* or *Nippostrongillus brasiliensis*. In some parasitic infection, the binding of parasite bound IgE antibodies to eosinophils induces the release of substances toxic for the parasites.[10] The main effector functions of IgE are therefore to be the mediators of allergy and defense against parasites.

BIOLOGICAL FUNCTIONS OF ANTIBODIES

Antibodies are the central component of the immune system. They have in common with the T cell receptor for antigen the fact that they recognize, a myriad of antigenic determinants in the universe. Both in the case of serum of membrane-bound antibodies and of T cells, this highly diverse repertoire ($>10^{10}$ putative possibilities in human or mouse) results from a complex genetic combinatorial process, unique to the immune system,[11,12]

For immunoglobulins, these genetic rearrangements occur in B lymphocytes prior to their encounter with antigen, producing,

Fig. 1.6. Structure of IgE.

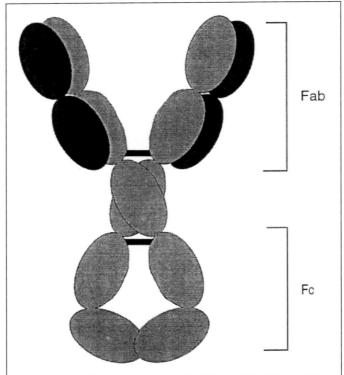

for each B cell clone, a unique receptor for antigen of IgM and IgD isotypes.

After a first exposure to an antigen, B cells with a specific receptor for that antigen are activated and, if the adequate accessory signals are provided, proliferate and differentiate into plasma cells producing essentially IgM antibodies. Iterative exposure to the same antigen induces a new set of genetic rearrangements in B cells, known as isotypic switching, which results in the production of IgG, IgA or IgE antibodies which keep the same V regions—and therefore the same antigen specificity—but exchange their μ heavy chain for a γ, α or ε heavy chain.[13]

The step that initiates the biological activities of antibodies is their combination with antigen. With the notable exception of rare catalytic antibodies,[14] the binding of an antibody does not modify its ligand. In fact, there is no persuasive evidence that antigen-antibody complexes exert any biological function by themselves.[15] To exert their numerous functions, antibodies interact with other proteins in the biological fluids and on cells. As illustrated in Figure 1.1, the antigen combining site of an antibody is formed by the association of the heavy and light chain variable region and is borne by the Fab region. No function other than binding antigen has been attributed to Fabs, nor to the light chain itself. The hinge region, of variable length (Figs. 1.3-1.6), has no other function than to allow, or restrict, flexibility of the antibody molecule. Although there is no biologically active site in the hinge region, the variable flexibility or rigidity may influence the binding to antigen and to effector molecules and subsequently modify antibody function.[16] It is through their Fc portion that antibodies exert their effector and regulatory function. Since the Fc portions are formed by the COOH-terminal parts of the immunoglobulin heavy chains, the sequence of the latter determines the potential activity of an immunoglobulin. The capacity to bind complement in the fluid phase or to specific receptors, called Fc receptors (FcR), on various cells governs the biology and antibodies and is fully dependent on their isotype. It has often been said that antibodies are bifunctional molecules, their V genes encoding their antigen-binding site and their heavy chain C genes their bioactive sites.

The present book reviews our knowledge of the molecules and the molecular steps that mediate the cell-dependent activities of immunoglobulin isotypes. Since the capacity of cells to functionally interact with immunoglobulins depends on their own machinery but also on the Fc receptors they express, the following chapters will analyze the structures of Fc receptors for the different immunoglobulin isotypes, their expression on different cell types, the structural bases of the FcR-mediated functions of antibodies as well as the clinical aspects of FcR dysregulation. In this chapter, I will put in perspective subsequent steps of the immune response in which antibodies play a role without depicting the structures of the molecules with which they interact and analyzing the molecular and cellular mechanisms involved.

ANTIGEN PRESENTATION

Except for bacterial polymeric antigens with repetitive determinants which directly activate B lymphocytes, the initial step of an immune response is the uptake, processing and presentation of antigen to T lymphocytes.[17] This process occurs in specialized antigen-presenting cells, heterogeneous population of cells found in all sites of the body where an immune response can be initiated. The most efficient antigen presenting cells are the bone marrow-derived dendritic cells of the blood stream and of secondary lymphoid organs,[18,19] the Langerhans cells of the skin,[20] the nonhematopoietic follicular dendritic cells of the secondary lymphoid organs[21] (spleen, lymph nodes...) and the antigen-stimulated B lymphocytes.[21] Blood monocytes and tissue macrophages are poor presenting cells for native antigen but become highly efficient when antigen is complexed with antibodies.[22] In fact, with the exception of B cells which efficiently capture antigen via their high affinity membrane immunoglobulin, the other presenting cells have no specific receptor for antigen. However, they are equipped with Fc receptors, at least for IgG and sometimes IgE, and very efficiently bind, ingest and process antigen complexed with IgG antibodies (Fig. 1.7). B lymphocytes, although also expressing Fcγ receptors, are unable to endocytose antigen-IgG complexes, preventing promiscuous antigen presentation.[23] Thus, in most situations, B cells only present antigens to which they have a specific immunoglobulin receptor. A notable exception to this rule is the IgE antibody which, when complexed to antigen, binds to Fcε receptors inducing endocytosis and subsequent presentation of antigen[24] (Fig. 1.7). This property of Ag-IgE complexes explains how allergens very efficiently keep up an IgE response, via a positive feed-back loop that aggravates the disease status.

The interaction of antigen-antibodies complexes, at least those formed by IgG and IgE with Fc receptors present at the surface of most antigen-presenting cells potentiates antigen presentation, boosting the specific immune response. It provides an explanation for the higher and more rapid secondary responses, when antibodies are already present in the body.

PHAGOCYTOSIS

Antigen presentation requires internalization of native antigen in the form of immune complexes. In many cases, these complexes are in soluble form and the internalization phenomenon is then called endocytosis. Often, however, antigen is in particulate, solid, form. This is the case for microorganisms such as parasites or bacteria. An immune response against these microorganisms can either be directly induced by B cell activation via antigen and mitogen receptors or can result from an

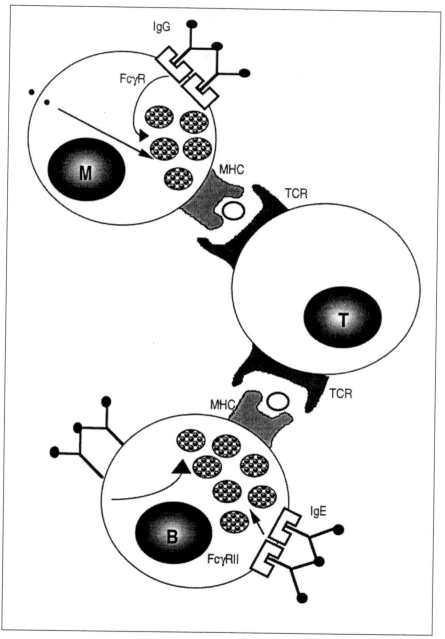

Fig. 1.7. Enhanced antigen presentation by Fc receptors for IgG on macrophages (M) and Fc receptors for IgE on B cells.

uptake of the microorganism by specialized cells, their cytoplasmic degradation and killing by cellular toxic compounds and the presentation of the antigens to T cells. The internalization of particulate antigens is called phagocytosis, and the cells that mediate them are called phagocytes.[25] Macrophages and neutrophils are the "professional" phagocytes, but many other cells, including mast cells,[26] can perform phagocytosis under certain conditions, when antigen is complexed to an antibody.

Antibodies, mostly IgG, but also IgE and perhaps IgA, play a crucial role in phagocytosis. Not only do they turn non-professional phagocytes into phagocytic cells,[26] but they also dramatically increase the efficacy of professional phagocytes for antibody-coated particles.[27-29] Since, in addition FcR triggering by antigen-antibody complexes stimulates the release of arachidonic acid,[30] nitric oxyde[31] and other oxygen free radicals[27,32] which are toxic for the ingested microorganisms, phagocytosis of the immune complex via Fc receptors provides the first line of defense against bacteria and parasites by killing invaders and initiating a specific immune response (Fig. 1.8).

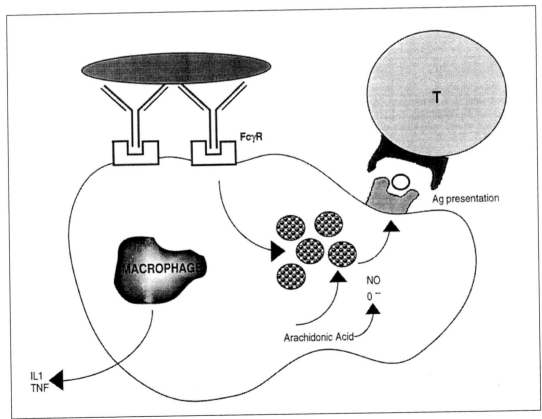

Fig. 1.8. Phagocytosis of immune complexes leading to macrophage activation and antigen presentation.

STIMULATION OF EFFECTOR FUNCTIONS: CYTOTOXICITY AND CYTOKINE RELEASE

T cells exert effector and regulatory functions by their antigen specific killing of target cells and the production of regulatory or effector cytokines. Different T cell subsets execute these functions. Cytotoxic T lymphocytes recognize antigen presented by major histocompatibility complex class I molecules on target cells and, after binding, inject toxic substances which kill their targets. Cytokines are mostly produced by two subsets of T helper cells, TH1 and TH2. Both are stimulated by antigen presented on major histocompatibility complex class II molecules. They differ by the pattern of the cytokines they produce. TH1 cells synthesize interleukin 2 (IL-2), Interferon-γ (IFN-γ) and tumor necrosis factor-α (TNF-α) which are responsible for cell-mediated immunity. TH2 cells synthesize IL-4, IL-5, IL-6, IL-10 and IL-13 which sustain antibody responses and therefore govern humoral immunity.[33] Other cytokines are produced by both TH1 and TH2 cells. Thus, TH1 cells are essential for defense against viruses, tumors and are responsible for graft rejection whereas TH2 cells are involved in allergic as well as anti-parasitic and anti-bacterial defences.[34]

The hallmark of T cell stimulation is its initiation via the T cell receptor for antigen. However, other cells can be induced to become cytotoxic and/or to produce cytokines. B lymphocytes, stimulated via their membrane immunoglobulins produce "TH2-like" cytokines, mostly IL-6 and IL-10.[35] Monocytes, macrophages, NK cells and mast cells can also be stimulated to become cytotoxic and/or produce cytokines. These cells lack antigen-receptors and are, in fact, activated by immune complexes via Fc receptors.

The interaction of monocytes and macrophages with antigen-antibody complexes not only provokes endocytosis, phagocytosis and an oxidative burst, but also the release of inflammatory cytoxins such as IL-1 and TNF[36,37] (Fig. 1.8) and the activation of their cytotoxic potential towards antibody-coated target cells.[38,39]

The binding of NK cells to IgG-sensitized target cells activates their lytic machinery resulting in target cell destruction and the release of "TH1-like" cytokines, such as IFN-γ and IL-2 that further sustain the anti-viral or anti-tumoral cytotoxic response[40] (Fig. 1.9).

Mast cells and basophils are triggered by IgE via high affinity IgE receptors to perform a series of activities. IgE is the physiological stimulus of mast cells. In contrast to almost all the other immunoglobulins which need to first bind antigen and become aggregated to have enough avidity to stably interact with Fc receptors, monomeric IgE binds to the high affinity mast cell, or basophil, Fcϵ receptor. This binding precedes the interaction with antigen but does not result in detectable signal. When multivalent antigen penetrates the body, it aggregates mast-cell associated IgE and their corresponding receptors, triggering cell activation.[41]

Upon activation, mast cells undergo morphological changes. They degranulate and release their granules containing arachidonic acid and metabolites, such as histamine, serotonin, platelet activating factor and leukotrienes[41] (Fig. 1.10). These substances are responsible for the vascular and muscular effects of allergic reactions, i.e., IgE is the molecule of allergy. IgE-complexes can also provoke the release of oxygen metabolites, leukotrienes, prostaglandins and platelet activating factor by monocytes, macrophages, platelets and polymorphonuclear cells,[39] reinforcing their role as the central element of allergic reactions.

Mast cells are not only effector cells for allergic reaction, they can also act as regulatory cells and produce cytokines. Again the interaction of antigen with mast cell bound IgE, which provokes an early release of inflammatory mediators, induces the late synthesis of cytokines, usually of the "TH2-type", such as IL-3 and IL-6 as well as TNF[42,43] (Fig. 1.11).

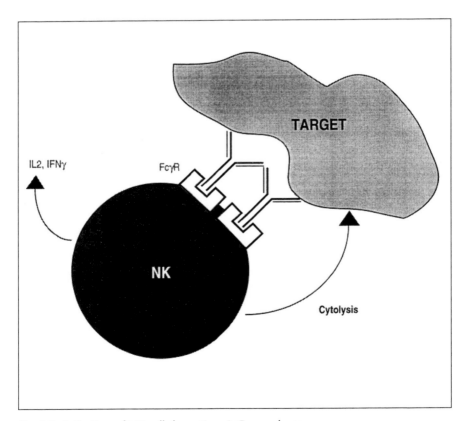

Fig. 1.9. Activation of NK cells by antigen-IgG complexes.

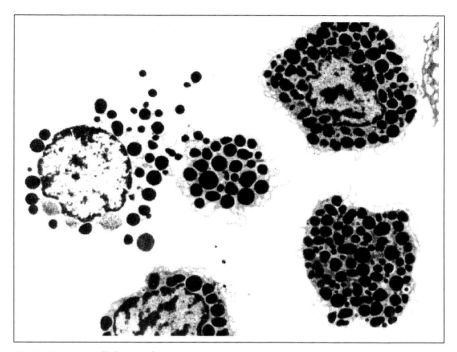

Fig. 1.10. Mast cell degranulation.

Through their activation with specific stimulatory Fc receptors on effector and regulatory cells, immunoglobulins play a pivotal role in the immune response, bringing antigen specificity to cells, other than T and B lymphocytes, which are not equipped with antigen-specific receptor. Moreover, immunoglobulin Fc-dependent interactions extend the concept of "TH1 and TH2 cells" to cells other than T cells, such as NK and mast cells (Table 1.2), involving them in a network of interactions that control the type, the magnitude and the length of immune responses.

NEGATIVE REGULATION OF IMMUNE RESPONSES

The administration of antigen to an animal stimulates its immune system and, if the conditions are fulfilled, the ultimate production of antibodies. If minute amounts of IgG antibodies are injected together with antigen, no antibody response occurs.[44] This phenomenon known as IgG negative feed-back requires intact IgG, F(ab')2 fragments having no effect, and is therefore Fcγ dependent.[45] Recent experiments have dissected the mechanism by which the IgG-negative feed-back functions. When antigen binds to membrane immunoglobulin on B lymphocytes, it aggregates it and initiates a series of events that lead to B cell proliferation and ultimate antibody production.[46] When IgG

Table 1.2. TH1 and TH2 subsets

	TH1	TH2	NK	Macrophage	Mast Cells
Stimulus	Ag	Ag	Ag-IgG	Ag-IgG	Ag-IgE
Interferon γ	++	-	++	-	-
Interleukin 2	++	-	++	-	-
Lymphotoxin (TNFβ)	++	-	++	-	-
GM-CSF	++	+	-	++	++
TNF α	++	+	-	-	++
Interleukin 3	++	++	-	-	++
Interleukin 4	-	++	-	-	++
Interleukin 5	-	++-	-	-	++
Interleukin 6	-	++	-	++	
Interleukin 10	-	++	-	++	
Interleukin 12				++	
Antibody Production					
IgM, IgG 1, IgA	+	++	+	+	-
IgG 2a	++	+	++	-	-
IgE	-	++	-	-	++
Cell Mediated Immunity	++	±	+	±	-

antibodies are fixed on the same antigen, their Fc portion binds to the Fcγ receptor expressed on the surface of B cells, and crosslinks it to the membrane immunoglobulin, which results in a negative signal for B cells[47] (Fig. 1.12). The molecular mechanisms of the FcγR-mediated negative signaling have been identified,[48-50] and the IgG negative feed back phenomenon has been extended to cells other than B lymphocytes. Thus, the crosslinking of Fcγ receptors with Fcε receptors on mast cells inhibits mast cell activation,[51] which provides a plausible mechanism for the sometimes successful desensitization procedures conducted in allergic patients by increasing the amount of IgG antibodies towards the allergen over that of IgE. Also, the crosslinking of Fcγ receptors to other activation receptors such as the T cell receptor, the NK cell Fcγ receptor, the Fcα receptor, or the monocyte/macrophage Fcγ receptors inhibits the activation of the corresponding cells, providing a central immunoregulatory role for IgG antibodies.[52]

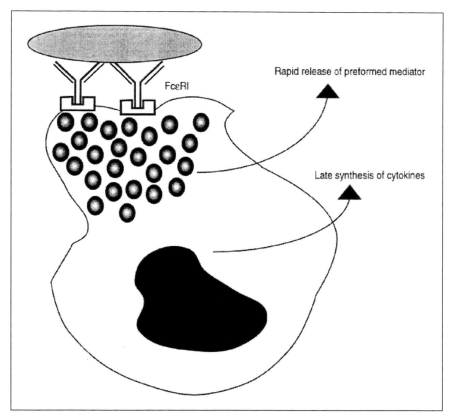

Fig. 1.11. Mast cell activation by IgE.

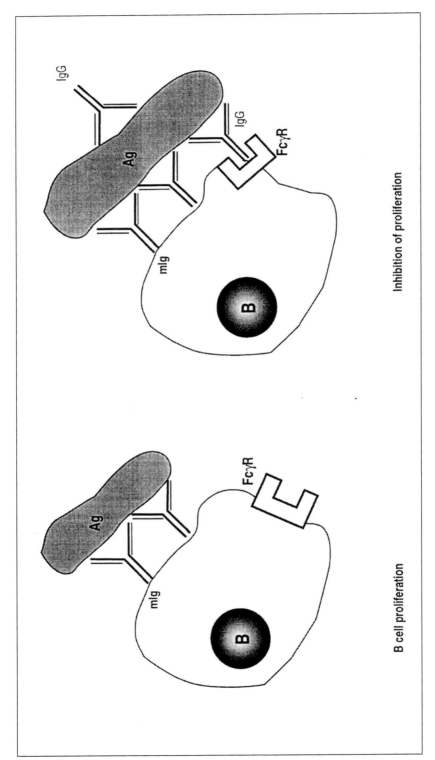

Fig. 1.12. Inhibition of B cell activation by Fcγ receptors.

TRANSPORT

To exert their effector or regulatory functions in certain parts of the body, like the intestinal tract, or to protect the immunologically immature fetus and infant, antibodies are transported across epithelia.

Mucosal transport

Antibodies that are secreted in mucosal tumor come from the lamina propria which is adjacent to the basolateral face of mucosal epithelial cells (Fig. 1.13). These antibodies, mainly IgA, cross the

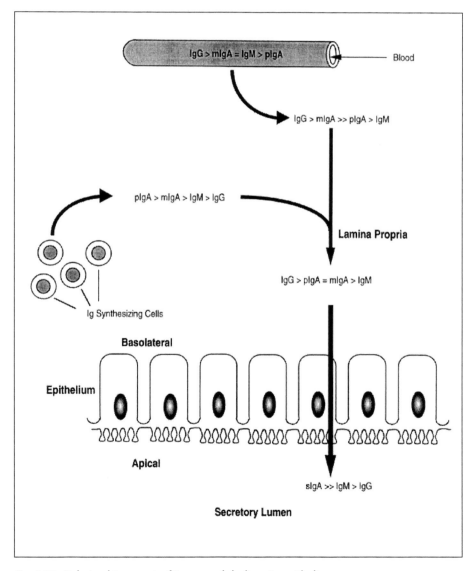

Fig. 1.13. Polarized transport of immunoglobulins via epithelia.

epithelium by a polarized active process with the help of the IgA-associated secretory component or after binding to a specialized Fc receptor called the poly-Ig receptor.[53] This phenomenon takes place in the intestine as well as at other mucosal membranes in the lungs or breasts.

Placental transport

The protection of the immunologically immature neonate is insured, for several weeks, by maternal antibodies which remain in the young child until his own immune system becomes efficient. In mammals, maternal IgG antibodies are transmitted to the fetus, before birth, through the placenta and IgA antibodies are transmitted to the suckling neonate through the milk. The transport of IgG antibodies through placenta is being currently studied and specific Fc receptors are being characterized. They need further analysis before a clear picture of their functions is drawn.

Protection of the neonate through the milk is a better known and complex phenomenon. The amount of maternal antibodies in suckling rats and mice rises during the first 12-19 days of life and then decreases. This is not due to the absence of antibodies in maternal milk but reflects a change in the permeability of the gut.

The presence of maternal antibodies is the consequence of a dual process: the secretion in the maternal milk, helped by the poly-Ig receptor, and the uptake by the neonatal small intestine by a neonatal Fc receptor whose expression is developmentally regulated.[54]

COMPLEMENT ACTIVATION

After interaction with antigen, IgM and most IgG antibodies bind the first component of complement and activate the classical pathway of the enzymatic cascade of the complement system. It is not within the scope of the present review, which is devoted to the cell-mediated functions of antibodies, to describe the complement system and the various ways by which it eventually destroys the bacterial walls. However, it should be kept in mind that complement activation produces a large series of active components which, via their interaction with specific receptors on many cells of the immune system exert direct biological activities or enhance antibody functions. The direct activities of complement include their role in inflammatory processes via anaphylatoxins generated by the cleavage of complement components (C3a and C5a) which activate and exert chemotaxis on neutrophils, monocytes or macrophages and increase vascular permeability and smooth muscle contraction. When bound on antigen-antibody complexes of IgM or IgG isotypes the activated third component of complement (C3) enhances phagocytosis of opsonized particles, endocytosis of antigens and their subsequent presentation to T cells as well as accessory B cell stimulation.[55]

Table 1.3. Biological activities of antibodies

Internalization	Cell Activation	Inhibition of Cell Activation
Phagocytosis	Release of preformed mediators (inflammation, cytotoxicity)	Prolilferation
Endocytosis	Synthesis of cytokines and of their receptors	Antibody production

CONCLUSION

Antibodies are a component of immune reactions, enhancing antigen presentation, regulating immune responses, triggering effector activities of complement and cells, leading to the elimination of antigen and maintenance of self integrity. By their bifunctional status they ensure specificity via their antigen-combining sites and promote efficacy and amplification by the binding of their Fc regions to a vast array of receptors on different cells, to exert multiple functions (Table 1.3).

REFERENCES

1. Silverstein AM. A History of Immunology. Academic Press, 1989.
2. Nezlin R. Immunoglobulin structure and function. In: Immunochemistry, Vand OSS CJ, Van Regenmortel MHV, eds. Marcel Dekker Inc. 1994; 3-45.
3. Cheng HL, Blattner FR, Fitzmaurice L et al. Structure of genes for membrane and secreted IgD heavy chains. Nature 1982; 296:410-15.
4. Rogers J, Early P, Carter C et al. Two mRNAs with different 3' ends encode membrane-bound and secreted forms of immunoglobulin μ chain. Cell 1980; 20:303-12.
5. Brack C, Hirama M, Lenhardt-Schuller R et al. A complete immunoglobulin gene is created by somatic recombination. Cell 1978; 15:1-12.
6. Williams AF, Barclay AN. The immunoglobulin superfamily-domains for cell surface recognition. Annu Rev Immunol 1988; 6:381-405.
7. Turner M. Antibodies and their receptors. In: Roitt I, Brostoff J, Male D, eds. Immunology. 4th ed. Mosby 1996; 4:01-11.
8. Burton DR. Antibody: The flexible adaptor molecule. Trends Biochem Sci 1990; 15:64-69.
9. Ishizaka T, Ishizaka K. Activation of mast cells for mediator release through IgE receptors. Prog Allergy 1984; 34:188-235.
10. Capron M, Grangette C, Torpier G et al. The second receptor for IgE in eosinophil effector function. Chem Immunol 1989; 47:128-178.
11. Alt FW, Blackwell TK, Yancopoulos GD. Development of the primary antibody repertoire. Science 1987; 238:1079-87.

12. Allison JP, Lanier LL. Structure, function and serology of the T cell antigen receptor complex. Annu Rev Immunol 1987; 5:503-40.

13. Esser C, Radbruch A. Immunoglobulin class switching: molecular and cellular analysis. Annu Rev Immunol 1990; 8:717-35.

14. Paul S, Volle DJ, Beach CM et al. Catalytic hydrolysis of vasoactive intestinal peptide by human autoantibody. Science 1989; 244:1158-62.

15. Metzger H. Fc Receptors and the Action of Antibodies. Am Soc Microbiol, Washington DC, 1990.

16. Burton DR. The conformation of antibodies. In: Fc Receptors and the Action of Antibodies. Metzger H. ed. Am. Soc Microbiol Wahington DC, 1990; 31-54.

17. Germain RN, Margulier DH. The biochemistry and cell biology of antigen processing and presentation. Annu Rev Immunol 1993; 11:403-50.

18. Steinman RM. The dendritic cell system and its role in immunogenicity. Annu Rev Immunol 1991; 9:271-96.

19. Inaba K, Metlay JP, Crowley MT et al. Dendritic cells as antigen presenting cells in vivo. Int Rev Immunol 1990; 6:117-26.

20. Esposito Farese ME, Sautes C, de la Salle H et al. Membrane and soluble FcγRII/III modulate The antigen-presenting capacity of murine dendritic epidermal Langerhans cells for IgG-complexed antigens. J Immunol 1995; 154:1725-36.

21. Line Y, Barthelemy C, de Boutheiller O et al. Memory B cells from human tonsil colonize mucosal epithelium and directly present antigen to T cells by rapid upregulation of B7.1 and B7.2. Immunity 1995; 2:239-48.

22. Manca F, Fenoglio D, Pira L et al. Effect of antigen/antibody ratio on macrophage uptake, processing and presentation to T cells of antigen complexed with polyclonal antibodies. J. Exp. Med. 1991; 173:37-48.

23. Lanzavecchia A. Receptor-mediated antigen uptake and its effect on antigen presentation to class II restricted T lymphocytes. Annu Rev Immunol 1990; 8:773, 93.

24. Kehry MR, Yamashita LC. Low affinity IgE receptor (CD23) function of B cells: role in IgE-dependent antigen focusing. Proc Nat Acad Sci 1989; 86:7556-60.

25. Metchnikoff E. Sur la lutte des cellules de l'organisme contre l'invasion des microbes. Ann. Inst. Pasteur 1887; 1:321-30.

26. Daëron M, Malbec O, Bonnerot C et al. Tyrosine-containing activation motif-dependent phagocytosis in mast cells. J Immunol 1994; 152:783-92.

27. Walker WS. Mediation of macrophage cytolytic and phagocytic activities by antibodies of different classes and class-specific Fc-Receptors. J Immunol 1977; 119:367-73.

28. Boltz-Nitulescu G, Langer K, Wilschke C et al. Mononuclear phagocytes express receptors for IgE which mediate phagocytosis of IgE-sensitized erythrocytes. Adv Biosci 1987; 66:113-21.

29. Anderson CL, Shen L, Eicher DM et al. Phagocytosis mediated by three distinct Fcγ receptor classes on human leukocytes. J Exp Med 1990; 171:1333-1446.

30. Rouzer CA, Scott WA, Hamill AL et al. IgE immune complexes stimulate arachidonic acid release by mouse peritoneal macrophages. Proc Nat Acad Sci 1982; 34:5656-60.

31. Dugas B, Mossalayi MD, Damais C, Kolb JP. Nitric oxide production by human monocytes: evidence from a role of CD23. Immunol Today 1995; 16:574-80.

32. Anderson CL, Guyre PM, Within JC et al. Monoclonal antibodies to Fc receptors for IgG on human mononuclear phagocytes: characterization and induction of superoxyde production by a monocyte cell line. J Biol Chem 1986; 261:12856-64.

33. Mosmann TR, Coffman RL. TH1 and TH2 cells: different patterns of lymphokine secretion lead to different functional properties. Annu Rev Immunol 1989; 7:145-73.

34. Peltz G. A role for CD4+ T cell subsets producing a selective pattern of lymphokines in the pathogenesis of human chronic inflammatory and allergic diseases. Immunol Rev 1991; 123:23-35.

35. Burdin N, van Kooten C, Galibert L et al. Endogenous IL6 and IL10 contribute to the differentiation of CD40-activated B lymphocytes. J Immunol 1995; 154:2533-44.

36. Mazingue C, Carriere V, Dessaint JP et al. Regulation of IgE synthesis by macrophages expressing Fcε receptors: role of Interleukin I. Clin Exp Immunol 1987; 67:587-93.

37. Latour S, Bonnerot C, Fridman WH, Daëron M. Induction of tumor necrosis factor-α production by mast cells via FcγR: role of the FcγRIII γ sub-unit. J Immunol 1992; 149:2155-62.

38. Fanger MW, Graziano RF, Shen L et al. FcγR in cytotoxicity exerted by mononuclear cells. Chem Immunol 1989; 47:214-53.

39. Dessaint JP, Capron M, Capron A. Immunoglobulin E release of mediators from mononuclear phagocytes, eosinophils and platelets, In: Fc Receptors and the action of antibodies, Metzger H, ed. Am Sc Microbiol, Washington DC, 1990; 260-87.

40. Anegon I, Cuturi MC, Trinchieri G et al. Interaction of Fc receptor (CD16) ligands induces the transcription of Interleukin 2 receptor (CD25) and lymphokine genes and expression of their production in human natural killer cells. J Exp Med 1988; 167:452-60.

41. Beaven MA, Metzger H. Signal transduction by Fc receptor: the FcεRI case. Immunol Today 1993; 14:222-226.

42. Plaut M, Pierce CW, Watson J et al. Mast cell lines produce lymphokines in response to cross-linkage of FcεRI or to calcium ionophores. Nature 1989; 339:64-67.

43. Fridman WH, Bonnerot C, Daëron M et al. Structural bases of Fcγ Receptor functions. Immunol Rev 1992; 125:49-76.

44. Henry C, Jerne NK. Competition of 19S and 7S antigen receptors in the regulation of the primary immune response. J Exp Med 1968; 127:133-40.

45. Sinclair NR, St C, Chan PL. Regulation of the immune response IV The role of the Fc fragment in feedback inhibition by antibody. In: Morpho-

logical and Fundamental aspects of immunity, Plenum Press, New York, 1971; 609-15.

46. Cambier JC, Pleinman CM, Clark MR. Signal transduction by the B cell antigen receptor and its co-receptors. Annu Rev Immunol 1994; 12:457-86.

47. Kölsch E, Oberbarnscheidt J, Brünner K, Heuer J. The Fc Receptor: its role in the transmission of differentiation signals. Immunol Rev. 1980; 49:61-78.

48. Amigorena S, Bonnerot C, Drake J et al. Cytoplasmic domain heterogeneity and functions of IgG Fc receptors in B lymphocytes. Science 1992; 256:1808-12.

49. Choquet D, Partisetti M, Amigorena S et al. Cross-linking of IgG receptors inhibits membrane Immunoglobulin-stimulated Calcium influx in B lymphocytes. J Cell Biol 1993; 121:355-63.

50. Muta T, Kurosaki T, Misulovin Z et al. A13-aminoacid motif in the cytoplasmic domain of FcγRIIB modultes B cell receptor signalling. Nature 1994; 368:70-3.

51. Daëron M, Malbec O, Latour S et al. Regulation of high-affinity IgE receptor mediated mast cell activation by murine low affinity IgG receptors. J Clin Invest 1995; 95:577-85.

52. Daëron M, Latour S, Malbec O et al. The same Tyrosine-based inhibition Motif, in the intracytoplasmic domain of FcγRIIB, regulates negatively BCR-,TCR-, and FcR-dependent cell activation. Immunity 1995; 3:635-646.

53. Underdown BJ. Transcytosis by the receptor for polymeric immunoglobulin. In: Fc Receptors and the action of antibodies, Metzger H, ed. Am Soc Microbiol, Washington DC, 1990; 74-93.

54. Simister NE. Transport of monomeric antibodies across epithelia. In: Fc Receptors and the action of antibodies, Metzger H, ed. Am Soc Microbiol, Washington DC, 1990:57-73.

55. Walport M. Complement, In: Immunology, Roitt I, Brostoff J, Male D, eds. Mosby, 4th edition 1996, 13:1-17.

CHAPTER 2

STRUCTURE AND EXPRESSION OF FC RECEPTORS (FCR)

Catherine Sautès

Different classes of cell surface receptors have the ability to interact with the Fc domain of Ig. The largest and most extensively characterized group are the FcR which belong to the immunoglobulin supergene family. These include receptors mediating major functions of the immune system and immunoglobulin transporters, exemplified by the poly-immunoglobulin receptor for IgM and IgA, and the IgG transporter of neonatal gut, recently characterized at the three-dimensional structure level. Some other FcR are lectin like molecules such as the low affinity receptor for IgE.

Since their initial discovery in the early 1970s, FcRs for all immunoglobulin isotypes have been described including IgG (FcγR), IgE (FcεR), IgA (FcαR), IgM (FcμR) and IgD (FcδR) on cells of the immune system. These cell surface glycoproteins mediate a variety of effector responses, when crosslinked by their ligands. Among the various FcR, FcγR and FcεR have been extensively studied. Most FcR are transmembrane glycoproteins with a ligand binding chain associated or not to one or two chains shared by several FcR and/or by other immune receptors. The extracellular domains of the ligand binding subunits consist of two to three Ig domains of the C2 family. They are highly conserved sequence identity values ranging from 70%-98% within the FcγR group to 40% between FcγR and FcεR. The FcγR mediate an exceptionally wide range of biological activities upon binding of IgG-antigen complexes. These include mediator release, internalization of complexes, antibody-dependent cell mediated cytotoxicity and negative regulation of other effector functions. Recent advances using mutagenesis approaches have led to the understanding of the structural

Cell-Mediated Effects of Immunoglobulins, edited by Wolf Herman Fridman and Catherine Sautès. © 1997 R.G. Landes Company.

basis of FcγR functional diversity. The common ligand-binding domains are coupled to distinct intracellular domains which transduce specific signals either directly or via the associated chains. Within these intracytoplasmic sequences, tyrosine activating motifs (immune tyrosine activating motifs or ITAM) shared by other cell surface receptors such as the BCR and the TCR have been identified as well as inhibitory motifs (immune tyrosine inhibiting motif or ITIM) involved in the negative regulation of other effector functions (see chapters 3 and 4).

In contrast with the FcγR family, the FcεRs comprise only two members. The high affinity FcεR, expressed by basophils, Langerhans cells and eosinophils, is responsible for the IgE-triggered allergic response. The low affinity FcεR which does not belong to the super Ig family but to an animal lectin family, has a much wider distribution on immune cells and plays a key role in B cell responses.

Soon after the discovery of FcγR, molecules binding antigen-complexed IgG were detected in culture supernatants of mouse lymphoid cells and were called immunoglobulin G-binding factors. These IgG binding factors (IgG-BF) were hypothesized to be derived from membrane FcγR. The existence of soluble forms of FcR was later generalized to isotypes other than IgG and to species other than mice. When the structures of FcR genes and protein have been elucidated it became apparent that these soluble receptors were either produced by proteolytic cleavage of the extracellular domain of FcR or encoded by alternatively spliced mRNA.

This chapter describes the structure and cellular distribution of the various FcR, with particular focus on FcγR and FcεR. The characteristics of the structure and functions of their soluble forms mentioned in this chapter are found in chapter 6.

NOMENCLATURE

The FcRs are designated by the subscript of the immunoglobulin isotype (Table 2.1). Thus, FcγR, FcεR, FcαR, FcμR and FcδR bind IgG, IgE, IgA, IgM, and IgD, respectively. Within FcR for a single immunoglobulin isotype, several classes have been defined. Class I receptors bind their ligand with high affinity. Three such receptors have now been characterized: FcγRI, FcεRI and FcαRI for monomeric IgG, IgE and IgA, respectively. Receptors that react only with antigen-complexed immunoglobulin are of class II or III (e.g., FcγRII, FcγRIII, FcεRII).[1]

Genes coding for FcR bear the name of the corresponding receptor. When several genes encode receptors of the same class, they bear the name of the receptor, followed by a capital Roman letter. For example, in humans, a total of eight genes have been described (Table 2.2). Three genes FcγRIA, FcγRIB and FcγRIC encode FcγRI, FcγRIIA, FcγRIIB and FcγRIIC genes encode FcγRII, and two genes FcγRIIIA and FcγRIIIB code for FcγRIII. Some genes have allelic forms, which are designated by capital Roman letters as superscripts. The human FcγRIIA gene has two alleles, FcγRIIA[HR] and FcγRIIA[LR], as does the human FcγRIIIB gene (FcγRIIIB^{NA-1} and FcγRIIIB^{NA-2}). Finally, alternative splicing of FcγR mRNA generates several transcripts of a single gene, designated by Roman letters and Arabic numbers. In the human two transcripts have been described for FcγRIB and FcγRIIA gene: FcγRIb1, FcγRIb2 and FcγRIIa1, FcγRIIa2 respectively. In man, three transcripts have been described for FcγRII, FcγRIIb1, FcγRIIb2 and FcγRIIb3, and four in mouse FcγRIIb, FcγRIIb2, FcγRIIb3 and FcγRIIb'1.

Table 2.1. FcR nomenclature

Species	Roman letter	m, h	(mFcR, hFcR)
Ligand isotype	Greek letter Subscript	γ, ε, α, μ, δ	FcγR, FcεR, FcαR...)
Classes	Roman numbers	I, II, III	(FcγRI, FcγRII...)
Genes	Roman capital letter	A, B, C	(FcγRIIA, FcγRIIIB...)
Alleles	Roman capital letter Superscript	LH/HR, NA1/NA2	(FcγRIIA^{LR}, FcγRIIIB^{NA-I}...)
RNA transcripts and proteins	Roman letter and arabic number	b1, b2, b3	(FcγRIIb1, FcγRIIb2...)
Subunit	Greek letter	α, β, γ	(FcεRIα, FcγRIIIAγ...)

Table 2.2. FcγR genes and isoforms

	Mouse			Human		
Genes (Alleles)	mFcγRI	mFcγRII (Ly 17.1,Ly17.2)	mFcγRIII	hFcγRIA	hFcγRIIA (HR/LR)	hFcγRIIIA
				hFcγRIB	hFcγRIIB	hFcγRIIIB (NA1/NA2)
				hFcγRIC	hFcγRIIC	
Chromosomal localization	3	1	1	1p13, 1q21	1q23	1q23
CD, Ly		Ly17		CD64	CD32	CD16
Receptor isoforms	mFcγRI	mFcγRIIb1, b'1,b2,b3*	mFcγRIII	hFcγRIa hFcγRIb1,b2 hFcγRIc	hFcγRIIa1,a2* hFcγRIIb1,b2,b3 hFcγRIIc	hFcγRIIIa hFcγRIIIb

*Soluble form, lacks transmembrane region

In addition, one must consider that some FcRs are composed of several subunits, sometimes shared by different FcR. Each subunit is designated by a Greek letter. For instance the high-affinity FcεRI is composed of three, α, β, γ subunits, γ being also part of the low-affinity FcγRIIIA, FcγRI and FcαRI.

Human FcRs are also defined by the CD nomenclature, CD64 for FcγRI, CD32 for FcγRII, CD16 for FcγRIII, and CD23 for FcεRII. Although convenient, this nomenclature completely excludes the large molecular heterogeneity of FcRs for a given isotype, which underlies their high functional diversity.

Table 2.3. Cell distribution of FcγR

Cells	Receptors										
	Mouse						Human				
	FcγRI	FcγRII			FcγRIII		FcγRIa	FcγRII		FcγRIII	
		b1	b'1	b2	b3			a1	a2 b1/b2/c	a	b
Neutrophils	+	?	?	+?	?	+	+	+	?	+	—	+
Eosinophils	—	?	?	?	?	?	+	+	?	?	—	+
Basophils	—	?	?	?	?	—	—	+	?	+	—	—
Mast cells	—	+	+	?	?	+	—	—	—	—	—	—
Monocytes	+	+	?	?	?	?	+	+	?	+	+	—
Macrophages	+	—	+	+	+	+	+	+	?	+	+	—
Langerhans cells	+	+	?	+	+	+	—	+	+	—	—	—
B cells	—	+	+	—	—	—	—	—	—	+	—	—
T cells	—	+	+	—	—	?	—	?	?	?	+	—
											TγCD3CD16+	
NK cells	—	—	?	—	—	+	—	±	—	—	+	—
Platelets	—	+	?	?	?	?	—	+	+	?	—	—

MEMBERS OF THE IMMUNOGLOBULIN SUPERFAMILY

RECEPTORS FOR IgG (FcγR)

In mice and humans three types of receptors bind IgG (Table 2.2).[1-7] FcγRI are receptors for monomeric IgG whereas FcγRII and FcγRIII bind IgG-containing immune complexes. The expression of FcγRI is restricted to cells of the monocyte/macrophage lineage whereas FcγRII are widely distributed on cells of the immune system (Table 2.3). FcγRII are found on lymphocytes, macrophages, platelets, polymorphonuclear cells, mast cells, and Langerhans cells. FcγRIII is the only FcγR expressed by all NK cells, and it is coexpressed with other FcγR on macrophages, basophils, neutrophils, and Langerhans cells. A significant structural homology, characteristic of the immunoglobulin supergene family, defines the FcγR family.

In mice, one gene encodes FcγRI, which is composed of three immunoglobulin-like ectodomains, a transmembrane spanning region, and an intracytoplasmic tail. A single gene encodes three transmembrane FcγRII glycoproteins, generated by alternative splicing: FcγRIIb1, FcγRIIb'1 and FcγRIIb2. FcγRIIb3 is a soluble receptor lacking the TM region of FcγRIIb2. One gene encodes the IgG-binding α chain of FcγRIII, which has 95% amino acid sequence homology with FcγRII in its ectodomains. It differs from FcγRII by its transmembrane and intracellular regions and because its expression requires association with an homodimer of γ chains.

In humans, three genes encode FcγRI isoforms, but only one transmembrane glycoprotein, homologous to the murine FcγRI, has been identified. Three genes encode FcγRII, where additional protein diversity is provided by alternative splicing of the first intracellular exon generating two transmembrane isoforms of FcγRIIB, FcγRIIb1 and FcγRIIb2. Finally, two genes encode two isoforms of FcγRIII, FcγRIIIA being homologous to its murine counterpart, with a transmembrane α chain associated to signal transducing chains. On neutrophils, FcγRIII-B is a phosphatidyl-inositol-anchored molecule.

The genes encoding the human and mouse FcγR are located on chromosome 1 with the exception of the mouse FcγRI on chromosome 3. These genes have arisen by duplication and divergence of a common ancestor gene.

The high affinity FcγR (FcγRI)

As compared to the other FcγR expressed by cells of the immune system (FcγRII and FcγRIII), FcγRI (Table 2.4) have two unique characteristics: they are high affinity receptors binding monomeric IgG and their extracellular region comprises three extracellular domains. The third extracellular domain is distinct from the first two domains shared by the other two FcγR, suggesting that the unique IgG binding characteristic of FcγRI are conferred by this additional domain.

In humans as in mice, FcγRI is expressed by cells of the monocyte macrophage lineage. It is also present on human neutrophils.

Mouse FcγRI

Mouse FcγRI is a 72 kDa glycoprotein with a 42 kDa polypeptidic backbone comprising an extracellular region of 273 amino acids containing three Ig-like domains, a single transmembrane region of 23

Table 2.4. Characteristics of FcγRI

	Mouse		Human			
Receptor isoforms	I		Ia	Ib2	Ib1	Ic
Molecular Mass (kDa)						
Glycoprotein	72		72	ND	ND	ND
Polypeptide	42		40			
Associated subunits	?		γ chain FcεRI	?		
Affinity for IgG (Ka)	10^7-10^8M^{-1}		10^8-10^9M^{-1}	<10^7 M^{-1}	ND	ND
Specificity for IgG						
Mouse	2a>>>1, 2b, 3		2a = 3>>>1,2b	ND	ND	ND
Human	3>1>4>>>2		3>1>4>>>2	ND	ND	ND
mAbs	—		10.1 32.2 197.1, 22, 44.1 62	—	—	—
Expression	Monocytes Macrophages		Monocytes Macrophages Neutrophils Eosinophils	ND	ND	ND
Regulation of expression	IFN-γ ⬈		IFN-γ ⬈ IL-10 ⬈ G-CSF ⬈ IL-4 ⬊	ND	ND	ND

amino acids and a cytoplasmic tail of 84 amino-acids. A single mFcγRI gene has been identified comprising six exons: two exons encoding the 5'-untranslated region (UTR) and leader sequences (L1 and L2), three exons encoding the extracellular region, one for each Ig-like domain (D1, D2, D3), and one exon encoding the transmembrane cytoplasmic and 3/UTR sequences. The gene is located on chromosome 3 which is syntenic with the region containing the genes on human chromosome 1 encoding hFcγRI. A polymorphism has been described for mFcγRI. The non-obese diabetic strain (NOD) express a receptor with a 73 amino acid deletion in the cytoplasmic tail due to a stop codon.

Mouse FcγRI binds monomeric IgG with high affinity ($Ka = 10^7-10^8 M^{-1}$). It is the only FcγR that binds a single mIgG subclass, IgG2a. Mouse FcγRI exhibits a specificity for human IgG1, IgG3 and to IgG4 with a lower affinity. Up to now, no mAbs have been obtained towards mFcγRI. This receptor has been detected on monocytes and macrophages and IFNγ increases its expression.

Human FcγRI

Of the three genes encoding hFcγRI, only FcγRIA encodes a membrane receptor, hFcγRIa. This receptor is a 72 kDa glycoprotein with a 40 kDa polypeptide backbone comprising an extracellular region of 292 amino acids with three Ig-like domains, a single transmembrane spanning region of 21 amino acids and an intracytoplasmic tail. Like its murine homologue FcγRI-A, it comprises six exons. The two other FcγRI genes, FcγRI-B an FcγRI-C have identical structure but contain stop codons in the exon encoding the third extracellular domain. Transcripts FcγRIb$_1$ and FcγRIc generated by FcγRIB and FcγRIC respectively, have been described but these may encode for soluble receptors which have not been identified yet. An additional transcript (FcγRIb$_2$) has been described which lacks sequences corresponding to D3, the third extracellular domain. Transfection experiments have shown that FcγRIb$_2$ binds only IgG complexes but not monomeric IgG. However no such protein has been identified on the membrane of hematopoietic cells. The three hFcγRI genes have been mapped at position 1q21. Recently, hFcγRIa has been shown to associate with homodimers of the γ subunit of the high affinity receptor for IgE, FcεRI. The complexes were detected on monocytes and could be reconstituted by cotransfection of the hFcγRIa and FcεRIγ subunit cDNA into COS cells. Comparison of cDNA and genetic sequences suggest genetic polymorphism of hFcγRIa. Individuals who lack FcγRI have been described in a Belgium family. However the structural basis of this polymorphism has not been determined yet.

Scatchard analysis of the binding of monomeric IgG have shown that hFcγRIa binds IgG with an affinity constant of $10^8-10^9 M^{-1}$. The receptor binds human IgG1 and IgG3 with high affinity and human IgG4 with a lower affinity but does not bind human IgG2. Notably, FcγRI bind aggregated IgG with similar affinity than monomeric IgG.

IFN-γ increases FcγRI expression on monocytes, macrophages and increases the expression of the receptor on neutrophils and on eosinophils. The promoter of hFcγRI gene contains an IFN-γ response element (GIRE) that binds STAT1 alpha transcription factor.[8] When isolated from human blood monocytes and neutrophils, FcγRI are saturated by serum IgG. This raises questions of how this receptor can distinguish IgG-opsonized particles. The IFNγ-induced upregulation of FcγRI which may occur at inflammatory sites may play a role in this function. G-CSF and GM-CSF induce also expression of FcγRI on neutrophils. Injections of GM-CSF are currently being used in cancer patients to increase FcγRI-dependent tumor cell toxicity exerted by neutrophils. On monocytes, whereas IL-10 upregulates hFcγRI expression, IL-4 was reported to downregulate its expression.

Several mAbs to hFcγRI have been described, e.g., 10.1, 197.1, 22, 62 and 44.1. Only two of them, 10.1 and 197.1 block the binding of IgG to FcγRI.

FcγRI–IgG interaction

By generating chimeric mFcγRI/mFcγRII receptors, the Ig binding roles of the extracellular domain of mFcγRI have been defined. The first two domains of mFcγRI bind IgG with low affinity and domain 3 confers high affinity binding to the receptor. Similar observations have been made in man. Notably, the unique binding of mIgG2a by mFcγRI appears to require interaction between domains 2 and 3 as replacement of domain 1 of mFcγRI by domain 1 of mFcγRII has no influence on binding, in contrast to the replacement of domains 1 and 2 of mFcγRI with those of mFcγRII. Thus domain 2 of mFcγRI seems to play a key role in IgG binding.

The Cγ2 domain of IgG is of importance for binding to FcγRI as shown by mAb inhibition studies and by the use of chimeric Ig generated between hIgG and mIgE where Cγ2 and/or Cγ3 were exchanged with Cε3 and Cε4. The region Leu234 to Ser239 (Leu-Leu-Gly-Gly-Pro-Ser) in Cγ2 which forms the hinge proximal region is crucial for the interaction. This sequence is present in all IgG isotypes that bind to hFcγRI with high affinity (hIgG1, hIgG3, mIgG2a, rat IgG2b and rabbit IgG) but diverges in mIgG2b and hIgG4 which bind to FcγRI with low affinity, mIgG2b containing Glu at position 235 and hIgG4 containing Phe at position 234 (Table 2.5). A second region of Cγ2 comprising a hinge proximal bend lying in proximity with the previous one is also implicated in binding, as shown by substitution of Pro331 situated in this loop. Finally, aglycosylation of Cγ2 decreases its affinity for hFcγRI.

In addition to Cγ2, studies with chimeric Ig have shown that Cγ3 does seem to play a role in the binding of IgG to FcγRI possibly by stabilization of the Fc structure.

The type 2 low affinity FcγR (FcγRII)

Among the various FcγR, FcγRII are the most diverse and widely distributed. Three isoforms encoded by a single gene have been described in the mouse and a total of six isoforms encoded by three distinct genes exist in humans. One of their characteristics is that some of these isoforms are soluble receptors, produced by alternative splicing. One of these isoforms has been described in mouse, FcγRIIb3 and one in the human, FcγRIIa2. The membrane isoforms are composed of two extracellular Ig-like domains, a transmembrane region and a cytoplasmic tail. They are found on virtually every FcγR-bearing cell with the exception of the NK cells. Unlike FcγRI, their affinity for monomeric IgG is low (Ka is less than $10^7 M^{-1}$) and they essentially interact with IgG immune complexes.

Mouse FcγRII

A single mFcγRII gene has been identified that encodes three transcripts generated by alternative splicing of transmembrane or intracytoplasmic sequences FcγRIIb1, b2 and b3 (Table 2.6). The mFcγRII gene comprises 10 exons, four encoding 5'UT and leader sequence (L1 to L3), two encoding the two extracellular domains (D1 and D2),

Table 2.5. Comparison of IgG C$_H$2 hinge proximal regions

Residue number*	Human				Mouse		
	IgG1	IgG2	IgG3	IgG4	IgG2a	IgG2b	IgG1
231	ALA	ALA	ALA	ALA	ALA	ALA	VAL
232	PRO	PRO	PRO	PRO	PRO	PRO	PRO
233	GLU	GLU	GLU	GLU	GLU	GLU	GLU
234	LEU	LEU	LEU	LEU	LEU	LEU	—
235	LEU	ALA	LEU	LEU	LEU	GLY	—
236	GLY	—	GLY	GLY	GLY	GLY	—
237	GLY	GLY	GLY	GLY	GLY	GLY	VAL
238	PRO	PRO	PRO	PRO	PRO	PRO	SER

* corresponding to human IgG

Table 2.6. Characteristics of mouse FcγRII

Receptor Isoforms*	IIb1	iib'₁	IIb2	IIb3
Associated subunits	—	—	—	—
Molecular Mass (kDa)				
Glycoprotein	40 - 60	40-60	40 - 60	50
Polypeptide	37	32	31	25
Affinity for IgG (Ka)	$<10^7$ M⁻¹	ND	ND	ND
Specificity for IgG				
mouse	1 = 2a = 2b>>>3	ND	1 = 2a = 2b>>>3	1 = 2a = 2b>>>3
human	3>1>2>4	ND	ND	ND
mAbs	2.4G2, anti Ly17.2	2.4G2	2.4G2	2.4G2
Cellular distribution	B cells Mast cells T cells (activated)	Macrophages B cells T cells Mast cells	Macrophages Langerhans cells Mast cells	Macrophages Langerhans cells

* a fourth isoform, FcgRIIb'₁, has been recently described. See text for details.

one the transmembrane region (TM) and three (C1 to C3) encoding the cytoplasmic and 3' UT sequences. The mFcγRIIb1 and mFcγRIIb2 transcripts of the mFcγRII gene encode transmembrane glycoproteins with a molecular weight ranging from 40-60 kDa. After removal of the N-linked saccharides, the apparent molecular weight of the receptors are 37 and 32 kDa respectively. They are composed of a 180 amino acid long extracellular region, a 26 amino acid transmembrane region and an intracytoplasmic tail of 94 amino acids for FcγRIIb1 and 47 amino acids for FcγRIIb2. Thus FcγRIIb2 differs from FcγRIIb1 only by the lack of a 47 amino acid sequence encoded by the first intracytoplasmic exon the receptor. FcγRIIb1 is mostly expressed in lymphocytes and mast cells whereas FcγRIIb2 is preponderant in macrophages and epidermal Langerhans cells.[9] A fourth membrane FcγRII isoform, FcγRIIb'1, has been identified recently at the mRNA and protein levels.[10] It differs from FcγRIIb1 by the lack of the C terminal 28 amino acids of the first intracytoplasmic exon. FcγRIIb'1 corresponds to a 32 kDa glycoprotein expressed in macrophages, B, T and mast cell lines, in normal spleen cells and in resting or LPS-activated B cells. Notably, the first IC exon of the human FcγRIIB gene ends by a single splice site in the downstream intron and encodes a 19 amino acid sequence homologous to the 19 residues of the first IC exon of FcγRIIb'1, suggesting that FcγRIIb'1 is the murine homologue of human FcγRIIb1. The third FcγRII isoform, FcγRIIb3, corresponds to a soluble receptor that circulates in biological fluids.[11] This 40-45 kDa glycoprotein gives, after deglycosylation, a 25 kDa polypeptide that comprises the extracellular region and the sequences encoded by the second and third intracytoplasmic exons, but lacks sequences encoded by the TM and the first intracytoplasmic exons. Macrophages and Langerhans cells secrete FcγRIIb3, as shown at the transcript level by PCR amplification and sequencing, and at the protein level by western blotting and characterization of CN-Br cleaved peptides.

The extracellular region of mouse FcγRII demonstrates an overall 60% amino acid identity with the hFcγRII extracellular counterpart. Mouse monoclonal antibodies have identified two allelic forms Ly17[a] and Ly17[b] that encode the two polymorphic forms of mFcγRII. The two alleles Ly17.1 and Ly17.2 differ in two codons encoding FcγRII, Pro116 and Gln161 being found in the Ly17.1 form and Leu116 and Leu161 in the Ly17.2 form. These amino acids are located in the second extracellular domain of the receptor. Notably, anti-Ly17 antibodies inhibit the binding of complexed IgG to the receptor, suggesting that these residues are located near the IgG-binding site.

Mouse FcγRII binds poorly monomeric IgG (Ka below $10^7 M^{-1}$). It also binds mouse IgE with low affinity. With the exception of mouse IgG3, mFcγRII bind all mouse IgG isotypes, IgG1, IgG2a and IgG2b. Studies of the association and dissociation kinetics between mouse IgG and mFcγRII reconstituted into planar membranes[12] have shown that mouse IgG1, IgG2a and IgG2b exhibit similar kinetic parameters, that mIgG3 does not bind and that possible allosteric changes that might occur in IgG1-anti DNP mAb after hapten binding do not appreciably affect the kinetic characteristics of mFcγRII binding. Low ionic strength buffer decreases the dissociation kinetics of mIgG to mFcγRII. Although a distinct receptor for mouse IgG3 has been detected by rosette formation with IgG3-sensitized erythrocytes, its molecular identification remains to be elucidated. Mouse FcγRII bind preferentially human IgG1 and IgG3 subclasses.

A number of reports have shown that aggregated IgG upregulates the expression of FcγR on T cells.[13,14] Mouse IgG1, IgG2b, IgG2a not IgG3 or the 2.4G2 MAb produce the effect. It is visible 6-12h after contact with ligand and can be detected qualitatively by rosette formation or quantitatively by ^{125}I-labeled 2.4G2 binding.[6] Whether IgG decreases FcγRII degradation or increases its expression has not been elucidated. Among cytokines, murine IFN-αβ was found to upregulate

FcγRII expression on hybridoma T cells[15] and recombinant IFN-γ to increase moderately FcγRII expression in mouse macrophages, due to the upregulation on FcγRIII and downregulation of FcγRII.[16] Finally, IL-4 decreases FcγRII expression in splenic B cells whereas it has no effect on B cells lines.[17]

Only two mAbs have been described that bind mFcγRII, 2.4G2, a rat mAb, the first mAb produced against FcγR and a mouse mAb anti-Ly 17.2 that detects the Ly-17.2 polymorphic form of mFcγRII. Both react with conformational epitopes and inhibit the binding of IgG into mFcγRII. The 2.4G2 mAb reacts with mFcγRII and mFcγRIII.

Human FcγRII

In contrast with the relatively simple murine situation, three hFcγRII genes have been described that encode a total of five distinct membrane receptors and one soluble receptor (Table 2.7). The three hFcγRII genes, FcγRIIA, hFcγRIIB and hFcγRIIC are similar in structure, each comprising eight exons. Two exons encode the 5'UTR and leader sequence (L1 and L2), two exons the extracellular region (D1 and D2), one the transmembrane region (TM) and three exons (C1, C2 and C3) encode the cytoplasmic domain and 3'UTR. The hFcγRII genes have been mapped to q23-24 on chromosome 1 and are linked with the hFcγRIII genes.

The hFcγRIIA gene encodes a membrane receptor (FcγRIIa1) and a soluble receptor (FcγRIIa2) produced by alternative splicing of the TM exon.[18] The hFcγRIIa1 glycoprotein has a molecular weight of 40 kDa in monocytes and was found to be associated with a homodimer of γ chains present also in FcεRI.[19] It resolves in a 36 kDa polypeptide after removal of the N-linked saccharides. The predicted receptor contains an extracellular region of 178 amino acids, a transmembrane region of 29 amino acids and cytoplasmic tail of 76 amino acids in Langerhans cells. The soluble FcγRIIa2 glycoprotein has an apparent molecular mass of 35 kDa in Langerhans cells, and the corresponding polypeptide of 32 kDa. The predicted receptor is identical to FcγRIIa1 in the extracellular and intracytoplasmic sequences but lacks the transmembrane region. The FcγRIIa transcripts have been detected by PCR in platelets, epidermal Langerhans cells,[20] as well as in some megacaryocytic, erythroleukemia and monocytic cell lines.[18, 21] The hFcγRIIa2 protein is found in serum.

The hFcγRIIB gene encodes three transcripts, hFcγRIIb1, hFcγRIIb2 and hFcγRIIb3. The FcγRIIb2 and b3 transcripts are produced by alternative splicing. The hFcγRIIb2 isoform is produced by alternative splicing of the first intracytoplasmic exon and FcγRIIb3 by alternative splicing of the second exon encoding the leader peptide. The predicted hFcγRIIb1 receptor is composed of an extracellular domain of 179 amino acids, a transmembrane region of 23 amino acids and a cytoplasmic tail of 61 amino acids. The predicted hFcγRIIb2 receptor is identical

Table 2.7. Characteristics of human FcγRII

Genes	hFcγRIIA		hFcγRIIB			hFcγRIIC
Receptor isoforms	**IIa1**	**IIa2**	**IIb1**	**IIb2**	**IIb3**	
Associated subunits	ND	ND	ND	ND	ND	ND
Molecular Mass (kDa) Glycoprotein	40	35	40			
Polypeptide	36	32	ND			
Affinity for IgG (Ka)	$<10^7 M^{-1}$	ND	$<10^7 M^{-1}$			$<10^7 M^{-1}$
Specificity for IgG Human	HR 3>1>>>2,4 (IIa1, IIa2) LR 3>1=2>>>4		3>1=4>>>2			
Mouse	HR 2a=1=2b (IIa1, IIa2) LR 2a=2b>>>1		2a=2b>>>1			
mAbs	IV.3, AT10, 41H16 ATIO, 41H16, (IIaHR), KB61, 2EI, KB61, KuFc79 CIKM5, KuFc79					
Cellular distribution	Monocytes (IIa1, IIa2) Macrophages Neutrophils Platelets (IIa1, IIa2) Langerhans cells (IIa1, IIa2)		Monocytes (IIb1,IIb2) Macrophages (IIb1,IIb2) B cells (IIb1,IIb2)			Monocytes Macrophages Neutrophils B cells NK cells
Regulation of expression	IL-4 ↓		ND			ND

except for the lack of the 19 amino acid sequence encoded by the first cytoplasmic exon. The hFcγRIIb1 and b2 transcripts have been found in B and myelomonocytic cells. The hFcγRIIb3 glycoprotein should be identical to hFcγRIIb2. The hFcγRIIb1 and hFcγRIIb2 glycoproteins have a molecular weight of 40 kDa. After removal of N-linked saccharides, the polypeptides have apparent molecular of 29 and 27 kDa respectively.

FcγRIIC results most likely from an unequal crossover event located in IC1 between FcγRIIA and FcγRIIB. The part of FcγRIIC upstream of this putative crossover point differs only in five nucleotides from the FcγRIIB gene and the part downstream of this point differs in only 19 nucleotides from FcγRIIA. The hFcγRIIC gene encodes a single transcript present in myelomonocytic cells, B cells and NK cells. The predicted protein has an extracellular region of 178 amino acids, a transmembrane domain of 29 amino acids and an intracytoplasmic tail of 75 amino acids. In vivo, the membrane expression of FcγRIIc remains to be established.

Notably, hFcγRIIA and hFcγRIIB encoded receptors have strong homologies in their respective extracellular and transmembrane regions (85% overall amino acid identity) but remarkably differ in their leader peptide and cytoplasmic tail.

Human FcγRII bind weakly monomeric IgG (Ka below $10^7 M^{-1}$) and react mostly with IgG-complexes. The heterogeneity of hFcγRII receptors raises the question of the isotypic specification of these distinct receptors. The two allelic forms, hFcγRIIa^{HR} and hFcγRIIa^{LR}, bind hIgG3, hIgG1 but not hIgG4. They markedly differ in the binding of human IgG2: the hFcγRIIa^{LR} exhibits strong binding and hFcγRIIa^{HR} a lower one. Both forms bind mIgG2a and mIgG2b, whereas only hFcγRIIa^{HR} binds mIgG1 strongly. Based on sequence comparisons, hFcγRIIb1, hFcγRIIb2, hFcγRIIb3 should bind IgG similarly since they have identical extracellular sequences. Transfection experiments have shown that hFcγRIIb1 binds human IgG1 and IgG3 >> human IgG2 > human IgG4 > and mouse IgG2a = mouse IgG2b > mouse IgG1.

Human FcγRII is widely distributed on cells of the immune system. It is expressed by almost all leukocytes with the exception of some T cells and most NK cells.[22] It is the only FcγR class present on basophils, platelets, B lymphocytes, Langerhans cells. It has also been detected on nonimmune cells such as trophoblasts and endothelial cells of the placenta. The products of the distinct hFcγRII genes are differentially expressed in some leukocytes. Whereas myelomonocytic cells express transcripts for all three FcγRII genes (a1, a2, b1, b2 and c), mature B cells express FcγRII-B gene products and some B cell lines express FcγRII-A. Megakaryocytic cell lines such as K562 express FcγRIIa1 and secrete FcγRIIa2. Cells and platelets also secrete FcγRIIa2.

Cytokines such as IFN-γ and IL-3 upregulate the expression of hFcγRII on eosinophils. IL-4 decreases their expression on monocytes.

Stimulation of B cells with anti-IgM or IL-4 results in significant increase in hFcγRIIb1 accompanied by a decrease in hFcγRIIb2 as shown by PCR.[23] Several mAbs have been produced which bind hFcγRII. These include IV.3, CIKM5, KuFc79, 41H16, ZE1, KB61, AT10, 7.30, 8.2, 8.26 and 8.7. With the exception of 8.2 and CIKM5 all these mAbs block the binding of IgG to FcγRII. Using chimeric FcR it has been shown that IV.3, 8.26, 8.7 and 7.30 react with epitopes located on the second extracellular domain whereas CIKM5 react with epitopes which involve the first and second extracellular domains.

The low affinity hFcγR genes and two of the genes for the IgE high affinity receptor (FcεRI α and γ subunit) are clustered on chromosome 1q23, as shown in Figure 2.1. This region is syntenic to mouse chromosome 1, where the FcγRII and FcγRIII mouse and human low affinity receptors genes are found. Comparison of FcγRII and FcγRIII gene organization and of cDNA sequences has suggested that FcγRII and FcγRIII derive from an ancestral FcR gene by duplication, recombination and diversification.[3] Mouse FcγRII would be the primordial receptors. The FcγRII evolutionary lineage involves the FcγRII genes whereas the FcγRIII lineage also includes the α chain of FcεRI. The hFcγRII A gene may have arisen by a recombination between mouse FcγRII and FcγRIII genes, whereas the 5' end of the gene has derived from FcγRIII and the 3' end from FcγRII. Comparison between the various hFcγRII genes with mFcγRII has suggested the following order for hFcγRII evolution: FcγRIIB → FcγRIIC → FcγRII-A. The FcγRI genes may have also derived from this pathway, via the addition of an exon encoding the third extracellular domain. The three hFcγRI also map to chromosome 1. The syntenic region on the mouse maps on chromosome 3.

Allelic forms of hFcγRIIA gene, hFcγRIIA[HR] and hFcγRIIA[LR] have been described. The monocytes of individuals expressing FcγRIIA[HR] (High Responders = HR) strongly stimulate T cells in the presence of anti-CD3 antibodies of mIgG1 isotype whereas those expressing FcγRIIA[LR] (Low Responders = LR) have a weak stimulatory capacity. Both alleles differ in two codons located in the first and second extracellular domains respectively. The two allelic variants have a glutamine or tryptophan at position 27 and an arginine or histidine at position 131 of the extracellular domains. They differ in their IgG binding capacity, the polymorphism at residue 131 being crucial for recognition of mouse IgG1 and human IgG2 whereas the polymorphism at residue 27 has no apparent effect. Transfection of native and chimeric cDNA have shown that arginine 131 is crucial for binding mouse IgG and histidine 131 for binding human IgG2. A polymorphism has also been described in hFcγRIIb1, tyrosine at position 235 being substituted by an aspartic acid. This substitution results in defective receptor internalization, as shown by transfection experiments.

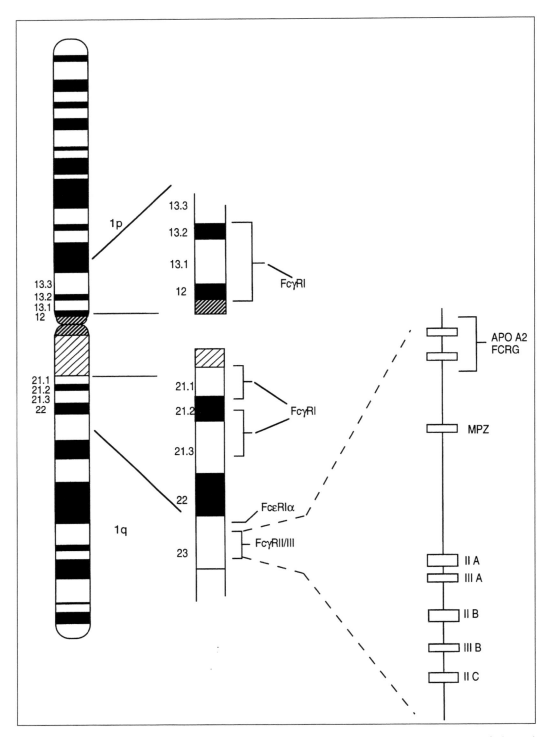

Fig. 2.1. FcR locus on human chromosome 1: FcγRI is distributed on 1p12-13 and 1q21. FcεRIα and γ(FCRG) subunits FcγRII and FcγRIII are present on 1q23.

FcγRII–IgG interaction

The identification and characterization of the HR/LR polymorphism of hFcγRII (on residue 131) and of the Ly17 polymorphism of mFcγRII (on residues 116 and 161) clearly implicate the second extracellular domain of FcγRII in the binding of IgG (Fig. 2.2). Studies using chimeric receptors between hFcγRII and hFcεRIα chain demonstrated that domain 2 is the IgG binding domain and that domain 1 of hFcγRII also seems to be involved. Its role seems most likely to influence receptor conformation, by stabilizing the structure of domain 2. Within the FcγRII domain 2, the IgG binding region is located on an eight amino acid segment (Asn154 to Ser161). Within this fragment, Ile155 and Gly156 are crucial for binding. Based on these studies and on the previously described related structure of CD4 domain, a three dimensional model of hFcγRII domain 2 has been proposed by Hogarth and colleagues.[7] The putative eight residue binding region is located in the F-G loop of domain 2 (Fig. 2.2), at the interface of domain 1. The hydrophobic residues at position 155 and 156 (Ile and Gly) may contribute to a hydrophobic cleft between the F-G and B-C loops. Notably Gly156 is conserved in all low affinity FcγR sequences and hydrophobic residues only are present at position 155 in all FcγRII sequences (Fig. 2.2). Other residues implicated by polymorphism studies in the binding of IgG by FcγRII are located in loop regions in close proximity to the 154-161 binding region = residue 131 of mFcγRII in the C'-E loop and residues Pro114 and Leu159 of hFcγRII in the adjacent B-C and F-G loops respectively.

Similarly to FcγRI, the Cγ2 region of IgG is the crucial binding site to FcγRII (Fig.2.3). Thus aglycosylation of the Cγ2 domain of hIgG1 and hIgG3 results in loss of binding. Sequence comparisons and studies using a panel of hIgG3 mutants in the 234-237 region have demonstrated that the Cγ2 domain binding site for hFcγRI and hFcγRII are similar, and correspond to the hinge proximal hydrophobic region, residues 234-237 (Leu Leu Gly Gly) (Table 2.5) being crucial for binding to both receptors. Notably whereas Leu234 is crucial for binding to hFcγRII, Leu235 seems important for binding to hFcγRI. In addition to Cγ2, Cγ3 may also play an important role by conferring another binding site or by stabilizing the Cγ2 structure.

Type 3 low affinity FcγR (FcγRIII)

Among the various FcγR, FcγRIII displays unique characteristics (Table 2.8). In the human, one of its isoform, FcγRIIIb, is not an integral membrane protein but is anchored to the plasma membrane via a glycosylphosphatidyl (GPI) anchor. Also, mFcγRIII, as well as the human isoform hFcγRIIIa, is the only FcγR expressed by NK cells. A soluble form of FcγRIII produced by proteolytic cleavage of membrane FcγRIII represents the most abundant soluble FcγR found in human blood. Notably it binds cell surface receptors other than Ig

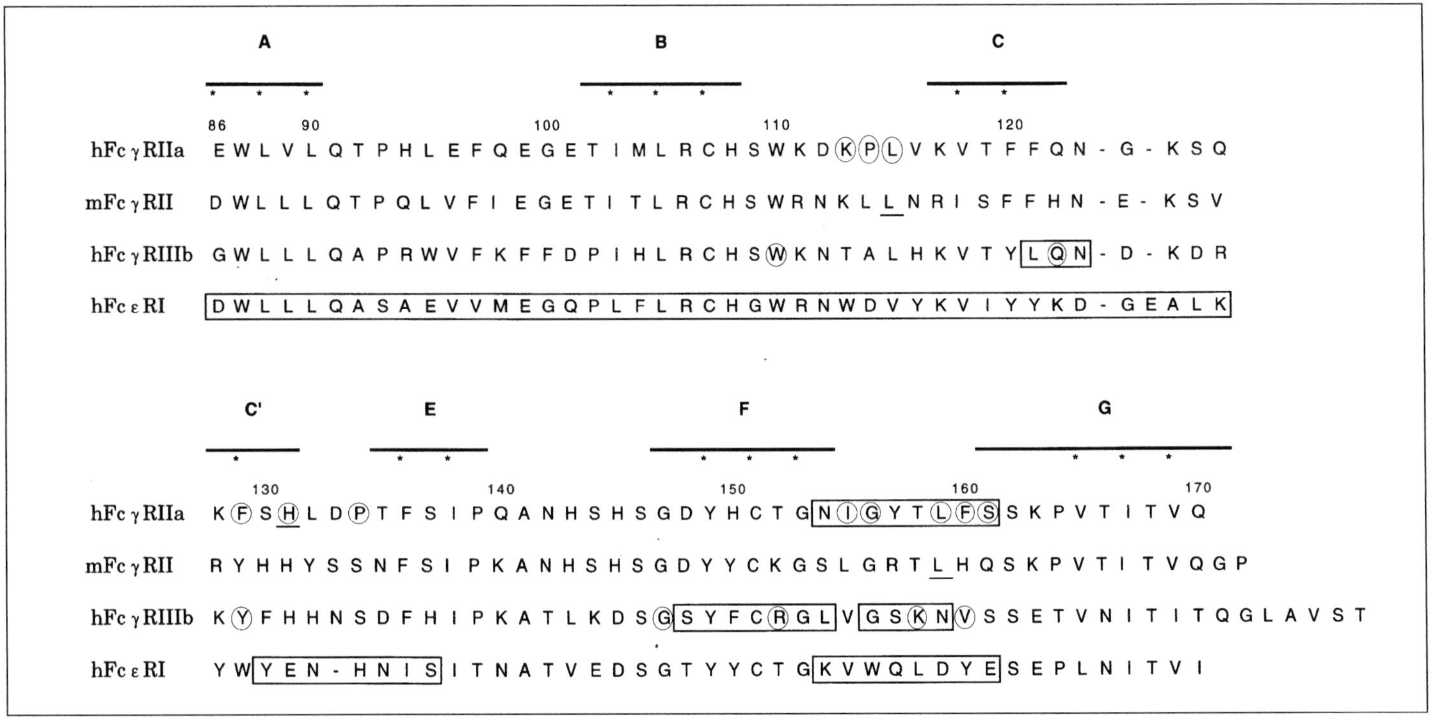

A

```
              86                    100
hFcγRIIa    E W L V L Q T P H L E F Q E G E T I M L R C H S W K D
mFcγRII     D W L L L Q T P Q L V F I E G E T I T L R C H S W R N K
hFcγRIIIb   G W L L L Q A P R W V F K F F D P I H L R C H S Ⓦ K N
hFcεRI     [D W L L L Q A S A E V V M E G Q P L F L R C H G W R N
```

B

```
                   110
hFcγRIIa    Ⓚ Ⓟ Ⓛ V K V T F F Q N
mFcγRII    L L N R I S F F H N
hFcγRIIIb   T A L H K V T Y L Ⓠ N
hFcεRI     W D V Y K V I Y Y K D
```

C

```
                120
hFcγRIIa    - G - K S Q
mFcγRII     - E - K S V
hFcγRIIIb   - D - K D R
hFcεRI     - G E A L K]
```

C'

```
              130
hFcγRIIa    K Ⓕ S Ⓗ L D Ⓟ T F S I P
mFcγRII     R Y H H Y S S N F S I P
hFcγRIIIb   K Ⓨ F H H N S D F H I P
hFcεRI     Y W [Y E N - H N I Ⓢ
```

E

```
                    140
hFcγRIIa    Q A N H S H S G D Y H C T G
mFcγRII     K A N H S H S G D Y Y C K G
hFcγRIIIb   K A T L K D S Ⓖ S Y F C Ⓡ
hFcεRI     I T N A T V E D S G T Y Y C T G
```

F

```
                     150              160
hFcγRIIa    Ⓝ Ⓘ G Y T L Ⓛ Ⓕ Ⓢ S K P V T I T V Q
mFcγRII     S L G R T L H Q S K P V T I T V Q G P
hFcγRIIIb   Ⓖ L V G S Ⓚ N Ⓥ S S E T V N I I T Q G L A V S T
hFcεRI     K V W Q L D Y E] S E P L N I T V I
```

G

```
                       170
```

Fig. 2.2. *Alignment of the second extracellular domain amino acid sequences of FcγR. The positions of the putative β strands are overlined. Regions implicated in the binding of IgG using chimeric receptor studies are boxed. Specific residues implicated using mutagenesis studies are circled. Polymorphic residues suggested to play a key role are underlined.*[7,36,37]

such as complement receptors type 3 and 4.[24, 25] and triggers cytokine production via these receptors. Thus FcγRIII can be viewed as a receptor for IgG and a membrane bound cytokine.

Mouse FcγRIII

In the mouse, FcγRIII is encoded by a unique gene linked to the mFcγRII gene on a genomic fragment located at the Ly-17 locus of chromosome 1. It comprises two exons encoding the 5'UTR and leader sequence, two exons for the extracellular region, and a single exon encoding the transmembrane and cytoplasmic regions and 3' UTR. Mouse FcγRIII is an integral membrane 40-50 kDa glycoprotein which resolves, after removal of N-linked oligosaccharides, into a 29 kDa

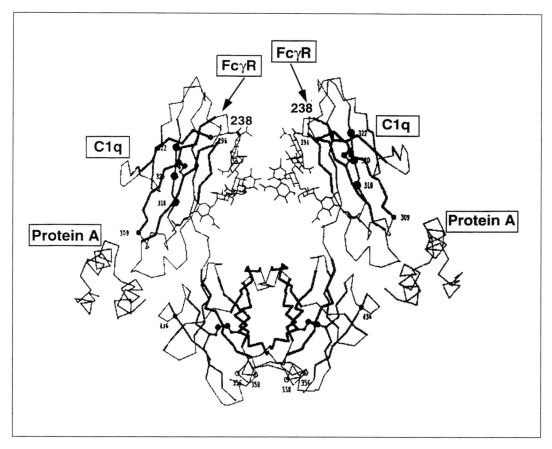

Fig. 2.3. Stereograph of the alpha-carbon backbone of human IgG1Fc using the coordinates defined by Deisenhofer (1981) showing binding sites for protein A, C1q (filled circles) and FcγR (residues 234 to 237 : Leu Leu Gly Gly). The beta pleated sheets are drawn with thick bonds, thicker for the first sheets. The Gm allotypic markers are indicated by open circles.

Table 2.8. Characteristics of FcγRIII

	Mouse	Human	
Genes	**FcγRIII**	**FcγRIIIA**	**FcγRIIIB**
Receptor isoforms			GPI
Associated subunits	αγ2,αβγ2	αγ2,αγζ,αζ2	
Molecular Mass (kDa) Glycoprotein Polypeptide	40 - 60 33	50 -80 33	50 - 80 29
Affinity for IgG (Ka)	$<10^7 M^{-1}$	$2 \times 10^7 M^{-1}$	$<10^7 M^{-1}$
Specificity for IgG mouse human	1 = 2a = 2b>>>3 3 = 1>2>>4	3>2a>2b>>1 3 = 1>>>2,4	3>2a>2b>>1 3 = 1>>>2,4
mAbs	2.4G2	CLBGran11 (NA1), MG38, ID3 GRM1 (NA2) 3G8, B73, VEP13, CLBGran1, BW209/2, Leu11a/b	
Cellular distribution	Macrophages NK cells Tγδ cells	Macrophages, NK cells Tγδ cells, Monocytes	Neutrophils Eosinophils
Regulation of expression	IFN-γ ↗	TGF-β ↗ (monocytes) IL-4 ↘	TNFα ↘ (neutrophils) IFNγ, GM-CSF, G-CSF ↗

polypeptide. It has a predicted extracellular region of 184 amino acids containing two Ig-like domains, a transmembrane region of 21 amino acids and a short cytoplasmic tail of 26 amino acids. Mouse FcγRIII has a strong sequence homology with mFcγRII in its extracellular domain (six amino acid differences) but diverges markedly from mFcγRII in the leader peptide, transmembrane and cytoplasmic tail.

The membrane expression of mFcγRIII requires coexpression of the γ chain of another FcR, the FcεRI high affinity receptor. On mast cells, immunoprecipitation experiments have shown that it is also associated to the β chain of FcγRI. The γ chain is a 8kDa polypeptide which comprises five amino acids in its extracellular region, a transmembrane region of 21 amino acids and a cytoplasmic tail of 42 amino acids. Notably, both in man and in mouse, the γ chain gene maps to chromosome 1, in the same region as the low affinity FcγR and the α chain of FcεRI (Fig. 2.1).

Natural killer cells and macrophages express FcγRIII. The only mAb reacting with mFcγR described so far, 2.4G2, detects both mFcγRIII and mFcγRII and inhibits IgG binding to these receptors.

HUMAN FCγRIII

Human FcγRIII glycoprotein exists in two isoforms, hFcγRIIIa and hFcγRIIIb that differ in membrane expression. Whereas hFcγRIIIa is a transmembrane molecule associated with the γ chain of FcεRI, hFcγRIIIb is anchored to the plasma membrane by a glycosyl phosphatidyl inositol (GPI) moiety. The hFcγRIIIa isoform is expressed in NK cells, monocytes, mast cells and FcγRIIIb in neutrophils. On mast cells it may be associated with the β subunit of FcεRI.[26] A cell type specific glycosylation of FcγRIIIa has been described, monocytic FcγRIIIa migrating as a broad band between 42 and 72 kDa and NK FcγRIIIa between 47 and 58 kDa.[27] FcγRIIIa and b give rise, after deglycosylation, to polypeptides of 33 and 29 kDa, respectively.

Two genes, hFcγRIIIA and hFcγRIIIB encode hFcγRIII, each producing a unique transcript, hFcγRIIIa and hFcγRIIIb. The two genes comprise two exons encoding the 5'UTR and leader sequence, one exon encoding each of the extracellular domains and one exon encoding transmembrane region, the cytoplasmic tail and the 3'UTR sequences. The genes are located on the q23,24 region of chromosome 1, linked to the hFcγRII genes (Fig. 2.1). The hFcγRIIIa and hFcγRIIIb transcripts encode receptors comprising an extracellular region of 190 amino acids, a transmembrane region of 21 amino acids and a cytoplasmic tail of 25 and 4 amino acids respectively. The shorter cytoplasmic tail of hFcγRIIIb results from a single nucleotide change generating a stop codon. Sequence analysis and mutagenesis experiments have shown that serine at position 203 plays a key role in the membrane anchoring of hFcγRIIIb via a GPI moiety. The hFcγRIIIa isoform contains a phenylalanine at this position which does not allow formation of the GPI anchor but instead the expression of a transmembrane molecule.

The hFcγRIIIa isoforms require association to small size transmembrane polypeptides for efficient cell surface expression. In NK cells hFcγRIIIa is associated with homo- and/or heterodimers of the FcεRIγ subunit and of the δ subunit of the T cell receptor linked by disulfide bridges. The genes encoding these subunits map to the q23.24 region of chromosome 1 that contains also the hFcγRII, hFcγRIII, FcεRI genes (Fig. 2.1). Transfection experiments have shown that hFcγRIIIa can additionally associate with the β subunit of FcεRI. Sequence comparisons of the transmembrane region of FcγRIII have shown the existence of a conserved stretch of eight amino acids, including an aspartic acid residue which may be crucial for association to the γ or δ subunits.

A polymorphism designated NA1/NA2 has been described for hFcγRIIIb. The NA2 form has six glycosylation sites and the NA1 has four. The two allelic variants differ in four amino acids. At two positions this results in the loss of N-linked glycosylation sites. The apparent molecular weights of hFcγRIIIb^{NA1} and hFcγRIIIb^{NA2} polypeptides slightly differ, that of hFcγRIIIb^{NA1} being 2.4 kDa smaller than that of hFcγRIIIb^{NA2}.

The two hFcγRIII isoforms bind human IgG with low affinity (around $10^7 M^{-1}$).

Human FcγRIIIb binds human IgG3 and IgG1 but not human IgG2 and IgG4. Binding of mouse IgG3, IgG2a and to a lesser extent mouse IgG1, but not mouse IgG2b, have been described for both isoforms.

Notably the extracellular domain of hFcγRIIIb binds other ligands than IgG. Neutrophil hFcγRIIIb interacts with complement receptor type 3 (CR3) or CD11b/CD18 and with FcγRII.[28-31] The interactions with CR3 occur via the lectin-like domain of CD11b and involve probably the high mannose-containing oligosaccharides of hFcγRIIIb.[32,33] Since hFcγRIIIb is a GPI-linked molecule, it is indeed unable to transduce by itself signals upon ligand binding. Its association to CR3 or to FcγRII may be crucial for signaling. Indeed the triggering of FcγRIIIb on neutrophils elicits signaling events which activate protein tyrosine kinases leading to phosphorylation of FcγRII.

The hFcγRIII isoforms differ in cell surface expression. Human FcγRIIIa is expressed on NK cells, macrophages a subpopulation of T cells (T γδ) cells and on monocytes, whereas hFcγRIIIb is exclusively present on neutrophils. The promoter regions of FcγRIII genes display tissue-specific transcriptional activities reflecting tissue distribution, and nucleotide differences that might contribute to cell type-specific transcription of FcγRIII genes have been identified.[34] The elements conferring the cell type-specific expression of the FcγRIII genes have been mapped recently within the 51 flanking sequences and first intron of the human FcγRIIIA and FcγRIIIB genes, by using transgenic mice.[35] Cytokines influence membrane levels of hFcγRIII. The expression of hFcγRIIIa is increased by TNF-β and hFcγRIIIb can be upregulated by IFNγ, TGF-β, GM-CSF and GCSF, and downregulated by TNFα. In addition IL-4 downregulates FcγRIII expression.

Several mAbs against hFcγRIII have been produced which include 3G8, 4F7, VEP13, Leu11a, b, and c B73.1, GRM-1, CLB Gran II, 1 D3 and BW209/2. The B73.1, Leu11c and CLB Gran 11 mAbs are specific for the NA1 form and GRM-1 for the NA2 form.

FcγRIII–IgG interaction

The two FcγRIII isoforms, hFcγRIIIa and hFcγRIIIb seem to have different affinities for hIgG (Ka = 2 x $10^7 M^{-1}$ and Ka < $10^7 M^{-1}$ respectively). The two NA1 and NA2 allelic forms which differ in glycosylation site number (4 and 6 respectively) may not interact similarly with IgG, the NA1 isoform binding better IgG3 than NA2. Moreover, the different glycosylated forms of FcγRIIIa found on NK cells with a high mannose oligosaccharide content and on monocytes with a low mannose content seem to have different affinities for IgG, FcγRIIIa on NK having higher affinity than FcγRIIIb on monocytes.

Like FcγRII, the second domain of FcγRIII is the IgG binding domain, as shown in the rat and in the human (Fig. 2.2). Residues in the F-G loop also play a major role. Other residues located in the B-C and C-C' loops may participate to the binding.[36,37]

The precise location of the FcR binding site on IgG is still not achieved. Both Cγ3 and Cγ2 domains seem important as well as the glycosylation of IgG.

It is of interest that the same structural elements on the low affinity Fcγ receptors, e.g., F-G loops, interact with the same region on the IgG molecules, e.g., the hinge region, between amino acids 233 and 237. Although the amino acid residues involved in ligand binding of FcγRIIIB are not conserved among Fc receptors, the F-G loop of different FcγR may constitute similar binding sites for IgG.

THE HIGH AFFINITY RECEPTOR FOR IGE (FCεRI)

The high affinity receptor for IgE, the receptor that initiates allergic reactions, has several major characteristic features (Table 2.9).[3,38,39] Although it binds monomeric IgE, it is totally unresponsive to the uncombined antibody but undergoes structural changes only when exposed to antigen-IgE complex. Interaction of receptor-bound IgE with a multimeric antigen is necessary to initiate cell activation. In addition FcεRI is a typical example of multimeric receptor containing three types of subunits, i.e., the α chain that contains the binding site for IgE, a β chain and two γ chains. Finally, affinity for ligand is very high, one to two orders of magnitude above that of FcγRI. This implies that only a few molecules of IgE needed to trigger activation

Table 2.9. Characteristics of FcεRI

Characteristic	Human FcεRI	Mouse FcεRI
Associated subunits	α β γ	α β γ
Receptor forms	α β γ2, α γ2	α β γ2
Molecular Mass (kDa)		
Glycoprotein	45-65(α), 32(β), 7-9(γ)	45-65(α), 32(β), 7-9(γ)
Polypeptide	26.4, 25.9, 7-8	25.8, 25.9, 7.8
Affinity for IgE	$10^{10}M^{-1}$	$10^{10}M^{-1}$
Chromosomal Localization	1q23(α), 11q13(β), 1q23(γ)	1q(α), 19(β), 1q(γ)
Cellular distribution	Mast cells Basophils Langerhans cells Eosinophils Monocytes (activated) Platelets	Mast cells Basophils

signals. FcεRI has been extensively studied in humans rats and mice.[3,38,39] Characteristics of FcεRI in these species will be discussed in parallel in view of their similarities.

The early characterization of FcεRI in the rat basophilic leukemia line RBL-2H3 led to the identification of a single polypeptide corresponding to the α chain. The initial failure to detect the additional β and γ subunits was due to the sensitivity of the noncovalent association between the subunits to mild detergents, purification of the intact complex requiring submicellar concentrations of detergent. In fact, FcεRI is a tetramer composed of a ligand-binding transmembrane glycoprotein, the α chain, a polypeptide that crosses the plasma membrane four times and has both amino and carboxy termini in the cytoplasm, the β-chain, and a dimer of two disulfide-linked-transmembrane polypeptides, the γ chains. A structural model for FcεRI has been proposed, based on sequences from the corresponding cDNA.[39] The α chain is a highly glycosylated polypeptide of molecular weight ranging from 45-65 kDa that resolves in a 26 kDa protein after treatment by N-glycosydases. The amino acid sequence deducted from cDNA shows that it is composed of an extracellular region of two Ig-like domains (180 amino acids in the human and 181 in the rats) a transmembrane region of

21 amino acids and cytoplasmic tails of 20, 25 and 31 residues for the rat, mouse and human α chains respectively. Sequence identity between the predicted human mouse and rat α chains is around 38%, the transmembrane region exhibiting the highest homology (62%). Interestingly a stretch of eight consecutive residues (LFAVDTGL) common to the transmembrane region of the three species α chains is also conserved in the transmembrane region of mouse and human FcγRIII. These residues may be involved in the interaction with the γ chains associated to these receptors. In addition, the two extracellular domains of the α chain are homologous both in size and sequence (35%) to the corresponding domains in FcγRIII. Among a total of 95 amino acids which are conserved in the rat mouse and human α chains, 61 are also found in both human and mouse FcγRIII, suggesting that the FcεRIα chain and FcγRIII genes probably arose from a common ancestor by gene duplication (see above). Notably FcεRIα and FcγRIII genes have a similar organization: two exons encode the 5'UTR and leader sequence, one exon each of Ig-like domains, and a single exon encoding the transmembrane, cytoplasmic tail and 3'UTR. The human FcεRIα gene has been mapped on chromosome 1, band 1q23 (Fig. 2.1), as the low affinity hFcγRI genes and its mouse homologue. The predicted rat mouse and human β-subunits comprise 243, 235 and 244 amino acids respectively with a high degree of sequence homology (69%).

Hydrophobicity plot analysis and mAb studies suggest that the FcεRIβ crosses four times the plasma membrane and has both the NH_2 and COOH termini in the cytoplasm. The FcεRIβ gene is a single gene composed of one exon for the 5'UTR and part of the NH_2 terminus, six exons for the transmembrane regions and one exon for the C terminus and 3'UTR. The mouse gene maps to chromosome 19, linked to the Ly-1 locus and the human gene to chromosome 11q13.

Comparison of sequences deducted from cDNA shows that the γ subunits from rat mouse and human receptor are highly conserved (86% amino acids identity). FcεRIγ polypeptide has a short extracellular region of five amino acids, a transmembrane region of 21 amino acids and a large cytoplasmic tail of 42 amino acids. There are two cysteine residues located at each end of the plasma membrane, the γ chain homodimers being formed via a disulfide bridge occurring between the N-terminal CySH residues. The FcεRIγ also associates with mFcγRIII, hFcγRIIIa, FcγRI and FcγRII and is also a member of the TCR-CD3 complex. The human γ chain subunit gene has been mapped to chromosome 1 band q23, the same region that contains the FcεRIα chain gene and the low-affinity FcγR genes. The mouse FcεRIγ has also been mapped on chromosome 1 to a region containing the FcεRIα gene and the low affinity FcγR genes. Transfection experiments have shown that in the rat and in the mouse both β and γ subunits are necessary for efficient surface expression of the α subunit. In contrast, in the human, coexpression of the γ chain is sufficient for expression of FcεRIα.

Monomeric IgE binds to FcεRI with high affinity ($Ka^{\sim}10^{10}M^{-1}$). Human FcεRI binds IgE from the three species whereas rodent FcεRI binds only rodent IgE. Similarly to FcγRIII, FcεRI binds lectins as recently shown by its capacity to bind εBP, a β-galactoside binding lectin previously identified as Mac2.

In rodents, FcεRI has been detected on two cell types only: mast cells and basophils. In the human it is also present on Langerhans cells, activated monocytes, eosinophils and was recently described in platelets at the mRNA and protein levels (M. Joseph et al, submitted). Several mAbs have been produced against rat and human FcεRIα chain. Epitope mapping studies have suggested that the second extracellular domain of the α chain plays a major role in IgE binding.

FcεRI–IgG interaction

FcεRI is a tetrameric complex composed of an IgE binding α subunit, a β subunit and a dimer of two γ subunits. The β and γ subunits have no direct role in the binding of IgE to FcεRI. As with FcγR, the second extracellular domain of hFcεRI is the IgE binding domain whereas the first domain plays a crucial role in maintaining the stability of the receptor to permit high affinity IgE binding. Rat IgE seems to also bind epitopes localized in domain 1 of FcεRI. Studies with human chimeric Fc receptors have shown that four regions contribute to the binding of IgE to domain 2: Ser93 to Phe104, Arg111 to Glu125, Tyr129 to Ser137 and Lys154 to Glu161. A molecular model of hFcεRIα domain 2 based on the structure of CD4 has been generated by Hogarth and colleagues. Loops located in the second domain and juxtaposed at the interface with domain 1 (F-G, C'-E and B-C loops) as well as B and C strands are involved in IgE binding (Fig. 2.2).

The principal domain in the Fc region of IgE involved in binding to FcεRI is Cε3. The roles of Cε2 and Cε4 are less clear, but they may provide supporting structural roles. Within Cε3, Asp330 to Leu363 region is necessary for binding. This may represent a similar structure to the hinge proximal region in IgG identified as crucial for the binding to FcγRI and FcγRII.

Receptor for IgA (FcαR)

The only FcαR described so far is the human myeloid high affinity FcαR (FcαRI) that belongs to the Ig superfamily and is structurally related to FcγR and FcεRI. Human FcαRI (CD89) is a 55 to 75 kDa integral membrane glycoprotein with a 32 kDa polypeptidic core which is associated with FcRγ chain.[40] The amino acid comprising sequence predicted from cDNA indicates that it is composed of 287 amino acids, an extracellular region of 206 amino acids, a transmembrane region of 19 amino acids and a cytoplasmic tail of 21 amino acids. The extracellular region is homologous to that of other FcR belonging to the Ig superfamily. Sequence comparisons suggest that

FcαR diverged from a common ancestor gene early in the evolution of the Ig superfamily FcR. It is not on the same chromosome as the other Ig superfamily members and has been mapped on chromosome 19q3.4. The gene for the human myeloid FcαR consists of five exons, two encode the leader peptide, two the two homologous Ig-like domains and the final exon encodes a short extracellular region, a hydrophobic transmembrane region and a short cytoplasmic tail.[41] Human FcαR is expressed associated with the γ chain of FcεRI. However transfection experiments have shown that the γ chain is not necessary for membrane expression of FcαR. The receptor binds IgA1 and IgA2 both in monomeric and polymeric forms, with an affinity for monomeric human IgA of $5 \times 10^7 M^{-1}$. It is expressed on monocytes, macrophages, neutrophils, eosinophils and mesangial cells. TGFβ downregulates its expression.[42] Several mAbs have been produced that recognize hFcαRI (My43, A3, A59, A62 and A77), My43 inhibiting the binding of IgA to the receptor.

POLY IGR

The poly IgR is the receptor that mediates transcytosis of polymeric IgG and IgA complex from the basolateral membrane of glandular epithelial cells to the apical membrane. After binding of the complexes to the poly IgR, complexes of ligand and receptors are endocytosed and migrate by transcytosis to the apical surface. The Ig complexes are then released in association with secretory component, a proteolytic cleavage product of the poly IgR. The poly IgR is a transmembrane glycoprotein with, depending on the species, three to five extracellular domains, a single transmembrane region and a cytoplasmic tail. The extracellular domains are highly conserved and exhibit a significant degree of homology with the variable domain of Ig, and are thus partially related to the leukocyte FcR which have homologies with the constant (C2) domains of Ig. Comparison of the predicted sequences of rabbit, rat and human poly IgR shows that the extracellular regions display 34% identity and the transmembrane and cytoplasmic regions 74 and 60% respectively, reflecting most probably conservation of sequences important for receptor function. The poly IgR is encoded by a single gene composed of a total of 11 exons. Several transcripts are produced by alternative splicing of the exons encoding the extracellular domains. A 14 amino acid peptide of the cytoplasmic tail appears crucial for transcytosis and the 30 C-terminal residues for rapid basolateral endocytosis of the receptor. The human poly IgR gene maps to chromosome 1q31-41. Cytokines such as IFN-γ, TNF-α and IL-4 increase expression of poly Ig-R on epithelial cells.

NEONATAL FcR (FcRn)

FcRn differs from the other FcR by several characteristic features. It is the receptor on intestinal epithelial cells which mediates the transfer of Ig from milk to the blood in newborn mice and rats. This receptor, which helps newborn animals to acquire passive immunity, is heterodimer composed of a transmembrane IgG-binding glycoprotein, the α chain, and of β2-microglobulin. The α chain of FcRn exhibits homologies with MHC class I antigens. The extracellular region of the α chain of FcRn comprises three Ig-like domains homologous to class I MHC antigens, a transmembrane region and a cytoplasmic tail. A three dimensional structure of a secreted form of FcRn has been obtained recently.[43,44] FcRn structure looks similar to the structure of MHC molecules (Fig. 2.4). The counterpart of the MHC peptide site is closed in FcRn making the FcRn groove incapable of binding peptides. A low resolution crystal structure of the complex between FcRn and Fc localized the binding site for Fc to a site distinct from the counterpart of the peptide groove.[43,44] FcRn binds to Fc at the interface between C_H2 and C_H3 domains (from Glu272 to His285 residues) which contains several histidine residues that could account for the pH-dependency of the FcRn/IgG interaction. FcRn co-crystallized as dimer with a single Fc. A model of interaction between Fc and FcRn at the cell surface has been suggested from these studies (Fig. 2.2).[43,44]

The gene encoding mouse FcRn has been mapped to chromosome 7, i.e., on a different chromosome than MHC antigens. The overall intron-exon organization of the FcRn gene is similar to that of the MHC class I gene. Phylogenetic tree analysis suggests that the FcRn gene diverged from MHC class I gene after the emergence of amphibians but before the split of placental and marsupial animals. FcRn bind IgG with high affinity (Ka 10^7, $10^8 M^{-1}$). Binding occurs only at low pH, in the acidic environment of the gut. The receptor is expressed in epithelial cells of the neonatal rat and mouse small intestine and in the yolk sac, as shown by northern blot. It is also present on the canalicular cell surface of adult hepatocytes suggesting that it may bind luminal IgG and provide a potential communication between parenchymal immune cells and bile.[45]

MEMBERS OF THE ANIMAL LECTIN SUPERFAMILY

THE LOW AFFINITY RECEPTOR FOR IgE (FcεRII)

The low affinity receptor for IgE, FcεRII, has several characteristic features. In contrast with the FcR described above, it displays a significant homology with a large family of calcium-dependent (C-type) animal lectins, including the asialo glycoprotein receptor and mannose binding proteins. Most probably because of this property human FcεRII binds to other molecules than IgE such as CD21, identified as the complement receptor 2 (CR2) and the receptor for Epstein-Barr virus, and with CD11b and CD11c, the α chains of the β2 integrin adhesion molecule complexes CD11b/CD18 and CD11c/CD18. FcεRII is a type II transmembrane glycoprotein, with a large external COOH-terminal region and a short NH_2-terminal intracytoplasmic tail. Similar to most FcR,

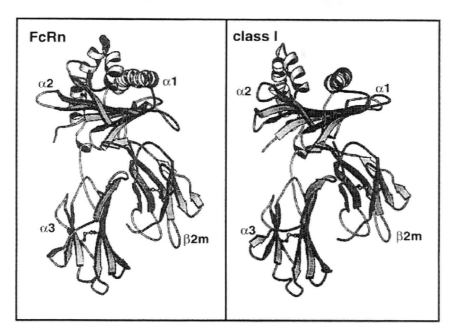

Fig. 2.4. (a, shown above) Ribbon diagram of FcRn. (Reprinted with permission from Burmeister et al. Nature 1994; 372:336-343, © MacMillan Magazines Ltd.) (b, shown below) Schematic view of the packing in the co-crystal to illustrate the possible network of 2:1 FcRn/Fc complexes forming on a membrane. (Reprinted with permission from Burmeister et al. Nature 1994; 372:379-383, © MacMillan Magazines Ltd.)

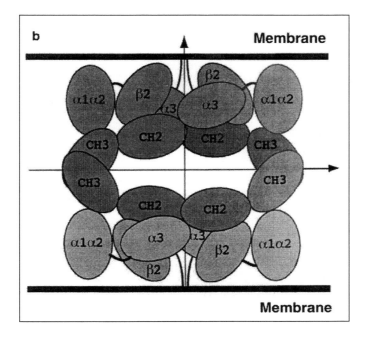

a soluble form of FcεRII capable of binding IgE and acting as a cytokine is released in the extracellular fluid. Thus, FcεRII could be viewed as a receptor for IgE, an adhesion molecule and a membrane bound cytokine.[46-49]

Structure

Human FcεRII (CD23) is a single chain 45 kDa glycoprotein. It contains one chain of N-linked carbohydrates of the complex type, several O-linked carbohydrates and sialic acid residues. The gene encoding FcεRII is present on chromosome 19, spans about 13kb, and consists of 11 exons. Exons 4-11 encode the external COOH terminal part of the molecule: exons 5-7 code for highly related 21 amino acids repetitive sequences, forming a leucine zipper, the closely linked exons 9-11 encode the soluble fragment released by cleavage of membrane FcεRII that binds IgE and is homologous to animal lectin family members. Notably the carbohydrate-binding domain of the asialoglycoprotein receptor shows exactly the same genomic arrangement as the FcεRII gene, suggesting a common ancestral origin of these two genes. The FcεRII promoter contains an IL-4 responsive element.

Two transcripts which differ by six amino acids at the cytoplasmic amino terminus are generated, FcεRIIa and FcεRIIb. FcεRIIb mRNA lacks the first two exons and starts with an optional exon that is located in intron II. FcεRII deducted polypeptide sequence is composed of a NH_2 terminal cytoplasmic region of 23 (FcεRIIa) or 18 (FcεRIIb) amino acids, a transmembrane region of 21 amino acids and an external 277 amino acid-long COOH region.

FcεRII binds IgE with intermediate affinity ($Ka = 10^7\text{-}10^8 M^{-1}$). Studies using a series of deletion mutants delimited the IgE binding site to a region of 128 amino acids (residues 160 to 287), which overlap the lectin domain. The region between the lectin domain and the membrane is predicted to form a α-helical coiled coil, called the stalk region, one of the main consequences of which is the formation of trimers on the cell surface.[50]

Proteolytic cleavage in the stalk region generates a major 25 kDa nonglycosylated polypeptide, expanding from amino acids 119 to 321. This fragment contains the IgE binding site as well as the homology domain to C type lectin family members, which spaces from Cys163 to Cys282, contains three intra-disulfide bridges, and, in reverse configuration, ("DGR"), the common recognition sequence of some integrin receptors. Chemical crosslinking studies have shown that this fragment forms hexamers in solution.[50]

The FcεRII binding site on human IgE has been mapped to the Cε3 constant region domain. Although human IgE is heavily glycosylated, and FcεRII a lectin-like structure, the IgE carbohydrates are not involved in the interaction. Notably, the binding sites of IgE to FcεRI and FcεRII are located close to each other.

Recent evidence indicates that FcεRII interacts with other molecules than IgE, namely the complement receptor 2 (CD21),[49,51] and the α chains of the β2 integrin heterodimers, CD11b/CD18 and CD11c/CD18.[52] The interactions with β_2 integrins are calcium dependent and inhibited by monosaccharides suggesting that they most probably involve the C-type lectin site of FcεRII and sugars present on the cellular ligands. The CD23-CD21 interaction may play a role in T-B cooperation for IgE production, and pairing of CD23 with CD11b or CD11c may be involved in inflammatory reactions.

Mouse FcεRII

The structure of mouse FcεRII is very similar to that found in humans. It is a 49 kDa glycoprotein containing two N-linked carbohydrates of the complex type, sialic acid residues and O-linked carbohydrates. After solubilization mouse FcεRII has the tendency to form oligomers. Murine FcεRII displays 74% homology with human FcεRIIa, but with several significant differences: (1) the presence of two, instead of one, glycosylation sites in the mouse receptor; (2) the murine FcεRII has a fourth 21 amino acid repetitive sequence; and (3) the murine FcεRII terminates just before the inverted RGD sequence, lacking this integrin receptor consensus motif. Mouse FcεRII also exists in two isoforms which are generated by alternative splicing and differ in their intracytoplasmic NH_2-terminus. Mouse FcεRII exist as an oligomer and as a monomer on the cell surface and analysis of engineered FcεRII molecules that had deletions in the stalk region indicated that formation of the oligomeric structure is dependent on this region.[53,54]

Soluble fragments are formed by proteolytic cleavage of surface FcεRII in the stalk region. Cleavage sites differ from those in hFcεRII. One consequence is the low ability of the mouse product to bind IgE consistent with a monomeric instead of oligomeric interaction with IgE.[55]

The two FcεRII isoforms have different cellular expression. FcεRIIa is found on resting B cells whereas FcεRIIb is expressed by the other Langerhans monocytes, macrophages, platelets and cells.

FcεRII is a B cell differentiation marker, restricted to mature B cells coexpressing sIgM and sIgD[56] and lost after Ig isotype switching as well as during the differentiation of B cells into Ig-secreting cells. Some transformed B cells, like the IgG-secreting RPMI8866 cells may express FcεRII after switching and, in some cases malignant B cells from patients with chronic-lymphocytic-leukemia may coexpress sIgG or sIgA together with CD23 even in the absence of sIgD.

FcεRII may also be found at low levels on a small proportion of normal T cells, or among T cells of patients with hyper IgE (Kimura disease). The expression of FcεRII was also detected on T cells from allergic donors stimulated in vitro with allergen. The constitutive expression of FcεRII by some human HTLV-1 transformed T cell clones has been demonstrated as well on T lymphocytes isolated from HIV-infected patients. Mice bearing an IgE-secreting plasmocytoma have a high proportion of their CD8+ T cells capable of binding IgE.[57] Whereas freshly isolated monocytes or alveolar macrophages bear little or no FcεRII, monocytes of patients with elevated IgE levels or alveolar macrophages from patients with allergic asthma express FcεRIIb.

Epidermal cells from patients with atopic dermatitis express FcεRII and release soluble FcεRII in the culture supernatants. Follicular dendritic cells, especially those localized in the light zone of germinal centers, are particularly rich in CD23 which may play a role in B cell differentiation through interaction with CD21. Recently FcεRII was also found on thymic epithelial cells, bone marrow stromal cells and keratinocytes. Eosinophils express FcεRII at variable levels in hypereosinophilic patients, as shown by immunohistochemistry, and both isoforms are detected by RT-PCR (M. Capron, in preparation).

Regulation of FcεRII expression
The expression of CD23 is regulated by cytokines and by surface molecules. On B cells, IL-4 triggers expression of both types of FcεRII[58] with a predominant effect on type b. The IL-4 signals leading to FcεRII gene activation are mediated via a PKC-dependent pathway. IL-13 which shares a common receptor with IL-4 was also shown recently to induce FcεRII expression on B cells. On the contrary, IFNγ, IFNα, TGFβ and TGFα inhibit IL4-induced expression of FcεRII on normal B cells. Down regulation by IFNγ and TGFα occur at the mRNA.[59] Whereas LTB4 and PAF increase the IL4-induced upregulation of FcεRII, PGE_2 inhibits this phenomenon.

Several B cell activation signals markedly increase the IL4-induced expression of FcεRII, such as contact-dependent interaction with T cells via CD40-CD40L pairing, engagement of sIg, or CD72. In contrast, mAbs against CD20 decrease FcεRII expression by stimulating the release of soluble FcεRII on the surface of EBV-transformed B cells and IL4-stimulated normal B cells.

The regulation of CD23 expression on monocytes is different from that on B cells. For example, freshly isolated normal monocytes do not express CD23 and, in contrast to B cells, IFN-γ does not suppress but rather enhances CD23 expression on monocytes. As in B cells, glucocorticoids and TGF-β inhibit CD23 on monocytes. The 1.25 dihydroxy-vitamin D3 inhibits CD23 expression on IL4-stimulated monocytes without affecting the expression of either the type a or b CD23 isoforms by highly-purified B cells. The regulation of CD23 on T cells appears to be similar to that on B cells.

Finally, the CD23 ligand, IgE was found to upregulate FcεRII in T and B lymphocytes.[6] As IgE itself was not found to increase the synthetic rate of FcεRII, the most likely explanation of these observation is that IgE occupancy of FcεRII decreases its degradation.

THE LOW AFFINITY RECEPTORS FOR IgM FOR IgD

Binding of IgM to human lymphocytes has been documented for more than two decades[60] using rosetting techniques. IgM binding to murine lymphocytes was detected more recently, using quantitative immunofluorescence techniques[57,61] and IgM binding to human NK cells has been similarly documented.[62] A 58 kDa PIPLC sensitive IgM binding molecule, upregulated by cell activation, was detected on B-lineage cells.[63]

Human T cells express also FcμR but with characteristics more distinct than those present on B cells: relative resistance to PIPLC treatment, downregulation upon cell activation and molecular size (60 kDa).[64]

Receptors for IgD (RFcδ) have been described on murine CD4+ T cells and T cell clones by rosetting techniques using IgD-SRBC.[65,66] Calcium ions are needed for optimal IgD-binding, as well as N-linked glycosylation, suggesting that RFcδ are animal lectin family members. Incubation of resting T cells with aggregated (but not monomeric) IgD or T cell activation causes RFcδ up regulation. The Fc μR and FcδR await further molecular characterization.

Fc RECEPTORS ON MICROORGANISMS

A fascinating evolutionary characteristic is the expression of molecules with high affinity and specificity for immunoglobulin by microorganisms. Besides their usefulness to purify IgG by protein A or G immunoadsorbents, their existence addresses the question of the role of these molecules in escape from immune attack and, more generally, in the microorganism's biology. Since this question has not been resolved, this chapter will only briefly summarize the structural and specificity knowledge of these receptors.

BACTERIAL Fc RECEPTORS

Six types of bacterial FcR are distinguished, according to their functional reactivity with different species and IgG subclasses.[67] Type I receptors, frequently termed protein A, are found on the surface of most strains of *Staphylococcus aureus*. They are composed of five homology units, each capable of binding Fc of IgG with a Ka of $3.10^6 M^{-1}$, and a sixth region that binds to cell walls. Protein A reacts with all human IgG subclasses. Type II receptors are also found on certain chains of group A streptococci. They display different IgG-binding specifications and await complete molecular characterization. Type III receptors are expressed on most group C and G streptococci. Also known as protein G, this receptor binds human IgG with high affinity ($Ka = 10^9 M^{-1}$) as well as most mammalian IgG. Its structure resembles that of protein A, with six domains, followed by a membrane-anchoring region. There is, however, no homology between the immunoglobulin-binding regions of protein A and protein G. Types IV, V and VI receptors, expressed on different streptococcal strains, bind IgG from different species with variable affinities. They have not yet been molecularly characterized. *Prevotella intermedia*, a suspected agent of adult chronic periodontitis display also IgG-Fc-binding activity. This FcR differs from the six types reported above and has a molecular mass of 65 kDa.[68]

Fc Receptors on Parasites

Some parasites are capable of reacting with immune complexes through specific receptors. For example, *Schistosoma mansoni* schistosomula bear a receptor reacting with the Fc portion of IgG and with β2-microglobulin. The interaction of immune complexes with this receptor activates the production by the parasite of a proteolytic enzyme that cleaves the Fc portion of IgG, protecting the microorganism from the immune attack. A similar mode of escape could be used by *Leishmania* and trypanosomes. The latter have three receptors, with different specifications for immunoglobulin isotypes, the expression of which varies with the cell cycle, and these are also released as soluble molecules.[69]

Virus-Associated Fc Receptors

Cells infected with herpes viruses, such as herpes simplex, cytomegalovirus, or varicellazoster virus, acquire FcR at their membranes. These receptors are encoded by the viruses and not the infected cells, and bind the Fc portion of IgG. The S peplomer protein of three distinct antigenic subgroup of the Coronaviridae (mouse hepatitis virus, bovine coronavirus and porcine transmissible gastroenterites virus) also bind IgG-Fc and react with 2.4G mAb.[70] The role of viral FcR in virus discrimination is still hypothetical.

References

1. Anderson C, Looney R. Review: human leukocyte IgG Fc receptors. Immunol Today 1986; 7:264-266.
2. Unkeless JC, Scigliano E, Freedman V. Structure and function of human and murine receptors for IgG. Annu Rev Immunol 1988; 6:251-281.
3. Ravetch J, Kinet J. Fc Receptors. Annu Rev Immunol 1991; 9:457-492.
4. Fridman WH, Bonnerot C, Daëron M et al. Structural bases of Fcγ receptor functions. Immunol Rev 1992; 125:49-76.
5. van de Winkel J, Capel P. Human IgG Fc receptor heterogeneity: molecular aspects and clinical implications. Immunol Today 1993; 14:215-221.
6. Daëron M, Sautès C, Bonnerot C et al. Murine type II Fcγ receptors and IgG-Binding Factors. Chem. Immunol 1989; 47:21-78.
7. Hulett MD, Hogarth MP. Molecular basis of the Fc receptor function. Adv Immunol 1994; 57:1-127.
8. Paquette R, Minosa M, Verma M et al. An interferon gamma-activation sequence mediates the transcriptional regulation of the IgG Fc receptor type IC gene by interferon-gamma. Mo. Immunol 1995; 32:841-851.

9. Esposito-Farèse ME, Sautès C, de la Salle H et al. Membrane and soluble FcγRII/III modulate the antigen presenting capacity of murine dendritic epidermal Langerhans cells for IgG-complexed antigens. J Immunol 1995; 155:1725-1736.

10. Latour S, Fridman WH, Daëron M. Identification, molecular cloning, biological properties and tissue distribution of a novel isoform of murine low-affinity IgG Receptor homologous to human FcγRIIB1. J Immunol 1996; in press.

11. Tartour E, de la Salle H, de la Salle C et al. Identification, in mouse macrophages and in serum, of a soluble receptor for the Fc portion of IgG (FcγR) encoded by an alternatively spliced transcript of the FcγRII gene. Intern. Immunol 1993; 5:859-868.

12. Hsieh H, Thompson N. Dissociation Kinetics between a Mouse Fc receptor (FcγRII) and IgG: Measurement by total internal reflection with fluorescence photobleaching recovery. Biochemistry 1995; 34:12481-12488.

13. Daëron M, Yodoi J, Néauport-Sautès C et al. Receptors for Immunoglobulin Isotypes (FcR) on murine T cells. I. Multiple FcR on T lymphocytes and hybridoma T lymphocytes and hybridoma T cell clones. Eur J Immunol 1985; 15:662-667.

14. Löwy I, Brézin C, Néauport-Sautès C et al. Isotype regulation of antibody production: T cell hybrids can be selectively induced to produce subclass specific suppressive Immunoglobulin-Binding Factors. Proc Natl Acad Sci USA 1983; 80:2323-2327.

15. Fridman WH, Gresser I, Bandu MT et al. Interferon enhances the expression of Fcγ receptors. J Immunol 1980; 124:2436-2441.

16. Weinshank RL, Luster AD, Ravetch JV. Function and regulation of a murine macrophage-specific IgG Fc receptor, FcγR-α. J Exp Med 1988; 167:1909-1925.

17. Conrad DH, Waldschmidt TJ, Lee WT et al. Effect of B cell stimulatory factor-1 (interleukin 4) of Fcε and Fcγ receptor expression on murine B lymphocytes and B cell lines. J Immunol 1987; 139:2290-2296.

18. Astier A, de la Salle H, de la Salle C et al. Human epidermal Langerhans cells secrete a soluble receptor for IgG (FcγRII/CD32) that inhibits the binding of immune-complexes to FcγR+ cells. J Immunol 1994; 152:201-212.

19. Masuda M, Roos D. Association of all three types of FcγR (CD64, CD32, CD16) with a γ chain homodimer in cultured human monocytes. J Immunol 1993; 151:6382-6388.

20. Schmidt D, Hanau D, Bieber T et al. Human epidermal Langerhans cells express only the 40-kilodalton Fcγ receptor (FcRII). J Immunol 1990; 144:4284.

21. Cassel DL, Keller MA, Surrey S et al. Differential expression of FcγRIIA, FcγRIIB and FcγRIIC in hematopoietic cells: analysis of transcripts. Mol Immunol 1993; 30:451-460.

22. Metes D, Galatiuc C, Moldovan I et al. Expression and function of FcγRII on human natural killer cells. Nat Immun 1994; 13:289-300.

23. Sarmay G, Rozsnyay Z, Koncz G et al. The alternative splicing of human FcγRII mRNA is regulated by activation of B cells with mIgM cross-linking, interleukin-4, or phorbolester. Eur J Immunol 1995; 25:262-268.

24. Galon J, Bouchard C, Fridman WH et al. Ligands and biological activities of soluble Fcγ receptors. Immunol Letters 1995; 44:175-181.

25. Galon J, Gauchat JF, Mazières N et al. Soluble Fcγ Receptor type III (FcγRIII, CD16) triggers cell activation through interaction with complement receptors. J Immunol 1996; 107:1184-1192.

26. Kurosaki T, Gander I, Wirthmueller U et al. The β subunit of the FcεRI Is associated with the FcγRIII on mast cells. J Exp Med 1992; 175:447-460.

27. Edberg CJ, Barinsky M, Redecha PB et al. FcRIII expressed on cultured monocytes is a N-glycosylated transmembrane protein distinct from FcγRIII expressed on natural killer cells. J Immunol 1990; 144:4729.

28. Zhou M-J, Todd III RF, van de Winkel JGJ et al. Cocapping of the leukoadhesin molecules complement receptor type 3 and lymphocyte function-associated antigen-1 with Fcγ receptor III on human neutrophils. J Immunol 1993; 150:3030-3041.

29. Poo H, Krauss J, Mayo-Bond L et al. Interaction of Fc gamma receptor type IIIB with complement receptor type 3 in fibroblast transfectants: evidence from lateral diffusion and resonance energy transfer studies. J Mol Biol 1995; 247:597-612.

30. Zhou M, Brown E. CR3 (Mac-1, alpha M beta 2, CD11b/CD18) and Fc gamma RIII cooperate in generation of a neutrophil respiratory burst: requirement for Fc gamma RIII and tyrosine phosphorylation. J. Cell. Biol 1994; 125:1407.

31. Petty HR, Todd III RF. Receptor-receptor interactions of complement receptor type 3 in neutrophil membranes. J Leuk Biol 1993; 54:492-494.

32. Stöckl J, Majdic O, Pickl A et al. Granulocyte activation via a binding site near the C-terminal region of complement receptor type 3 α-chain (CD11b) potentially involved in intramembrane complex formation with glycosylphosphatidylinositol-anchored FcγRIIIB (CD16) molecules. J Immunol 1995; 154:5452.

33. Thornton B, Vetvicka V, Pitman M et al. Analysis of the sugar specificity and molecular location of the β-glucan-binding lectin site of complement receptor type 3 (CD11b/CD18). J Immunol 1996; 156:1235.

34. Gessner JE, Grussenmeyer T, Kolanus W et al. The human low affinity immunoglobulin G Fc Receptor III-A and III-B genes. J Biol Chem 1995; 270:1350-1361.

35. Li M, Wirthmueller U, Ravetch JV. Reconstitution of human FcγRIII cell type specificity in transgenic mice. J. Exp. Med 1996; 183:1259-1263.

36. Tamm A, Kister A, Nolte KU et al. The IgG binding site of human FcγRIIIB receptor involves CC' and FG loops of the membrane-proximal domain. J Biol Chem 1996; 271:1-8.

37. Hibbs M, Tolvanen M, Carpén O et al. Membrane-proximal Ig-like domain of FcγRIII (CD16) contains residues critical for ligand binding. J Immunol 1994; 152:4466-4474.

38. Metzger H, Alcaraz G, Hohman R et al. The receptor with high affinity for immunoglobulin E. Annu Rev Immunol 1986; 4:419-470.

39. Kinet JP, Metzger H. Genes, structure, and actions of the high-affinity Fc receptor for Immunoglobulin E. In: Metzger H, ed. Fc receptors and the action of antibodies. ASM, Washington, USA 1990; 239-259.

40. Morton H, Van den Herik-Oudijk I, Vossebeld P. Functionnal association between the human myeloid IgA Fc receptor (CD89) and FcRγ chain. J Biol Chem 1995; 270:29781-29787.

41. de Wit T, Morton H, Capel P et al. Structure of the gene for the human myeloid IgA Fc receptor (CD89). J Immunol 1995; 155:1201-1209.

42. Reterink TJ, Levarht EW, Klar-Mohamad N et al. Transforming growth factor-beta 1 (TGF-beta 1) downregulates IgA Fc-receptor (CD89) expression on human monocytes. Clin Exp Immunol 1996; 103:161-166.

43. Burmeister W, Huber A, Bjorkman P. Crystal structure of the complex of rat neonatal Fc receptor with Fc. Nature 1994; 372:379-383.

44. Burmeister WP, Gastinel LN, Simister NE et al. Crystal structure at 2.2 A resolution of the MHC-related neonatal Fc receptor. Nature 1994; 372:336-343.

45. Blumberg R, Koss T, Story C et al. A major histocompatibility complex class I-related Fc receptor for IgG on rat hepatocytes. J Clin Invest 1995; 95:2397-2402.

46. Conrad DH. FcεRII/CD23: The low affinity receptor for IgE. Annu Rev Immunol 1990; 8:623.

47. Delespesse G, Suter U, Mossalayi D et al. Expression, structure and function of the CD23 antigen. Adv Immunol 1991; 49:149-191.

48. Delespesse G, Sarfati M, Wu CY et al. The low affinity receptor for IgE. Immunol Rev 1992; 125:77-97.

49. Bonnefoy J-Y, Lecoanet-Henchoz S, Aubry JP et al. CD23 and B cell activation. Current Opin Immunol 1995; 7:355-359.

50. Beavil R, Graber P, Aubonney N et al. CD23/FcεRII and its soluble fragments can form oligomers on the cell surface and in solution. Immunology 1995; 84:202-206.

51. Aubry J-P, Pochon S, Graber P et al. CD21 is a ligand for CD23 and regulates IgE production. Nature 1992; 358:505-507.

52. Lecoanet-Henchoz S, Gauchat J, Aubry J et al. CD23 regulates monocyte activation through a novel interaction with the adhesion molecules CD11b-CD18 and CD11c-CD18. Immunity 1995; 3:119-125.

53. Beavil A, Edmeades R, Gould H et al. α-Helical coiled-coil stalks in the low-affinity receptor for IgE (FcεRII/CD23) and related C-type lectins. Proc Natl Acad Sci USA 1992; 89:753-760.

54. Dierks S, Bartlett W, Edmeades R et al. The oligomeric nature of the murine FcεRII/CD23: Implication for function. J Immunol 1993; 150:2372.

55. Bartlett WC, Kelly A, Johnson C et al. Analysis of murine soluble FcεRII sites of cleavage and requirements for dual-affinity interaction with IgE. J Immunol 1995; 154:4240-4246.

56. Waldschmidt TJ, Conrad DH, Lynch RG. The expression of B cell surface receptors.I. The ontogeny and distribution of the murine B cell IgE Fc receptor. J Immunol 1988; 140:2148-2154.

57. Mathur A, Lynch R, Köhler G. The contribution of constant region domains to the binding of murine IgM to Fcµ receptors. J Immunol 1988; 140:143.

58. Waldschmidt TJ, Conrad DH, Lynch RG. The expression of B cell surface receptors.II. Interleukin 4 can accelerate the developmental expression of the murine B cell IgE Fc receptor. J Immunol 1989; 143: 2820-2827.

59. Hashimoto S, Koh K, Tomita Y et al. TNF-α regulates IL-4 induced FcεRII/CD23 gene expression and soluble FcεRII release by human monocytes. Int Immunol 1995; 7:705-713.

60. Moretta L, Ferrrarini M, Durante ML et al. Expression of a receptor for IgM by human T cells in vitro. Eur J Immunol 1975; 5:565.

61. Mathur A, Lynch RG, Kohler G. Expression, distribution and specificity of Fc receptors for IgM on murine B cells. J Immunol 1988; 141: 1855-1862.

62. Pricop L, Rabinowich H, Morel P et al. Characterization of the Fc mu receptor on human natural killer cells. J Immunol 1993; 151:3018-3029.

63. Sanders SK, Kubagawa H, Suzuki T et al. IgM binding protein expressed by activated B cells. J Immunol 1987; 139:188.

64. Nakamura T, Kubagawa H, Ohno T et al. Characterization of an IgM Fc-binding receptor on human T Ccells. J Immunol 1993; 151:6933-6941.

65. Swenson C, Amin A, Xue B et al. Regulation of IgD-receptor expression on murine T cells. I. Characterization and metabolic requirements of the process leading to their expression. Cell Immunol 1993; 152:405-421.

66. Amin A, Swenson C, Xue B et al. Regulation of IgD-receptor expression on murine T cells. Cell Immunol 1993; 152:422-439.

67. Burton DR, Woof JM. Human antibody effector function. Adv Immunol 1992; 51:1-84.

68. Labbe S, Grenier D. Characterization of the human immunoglobulin G Fc-binding activity in prevotella intermedia. Infect Immun 1995; 63:2785-2789.

69. Vincendeau P, Daëron M. *Trypanosoma musculi* co-express several receptors binding rodent IgM, IgE and IgG subclasses. J Immunol 1989; 142:1702-1709.

70. Oleszak E, Kuzmak J, Hogue B et al. Molecular mimicry between Fc receptor and S peplomer protein of mouse hepatitis virus, bovine corona virus, and transmissible gastroenteritis virus. Hybridoma 1995; 14:1-8.

CHAPTER 3

CELL ACTIVATION
VIA FC RECEPTORS

Christian Bonnerot

Unlike many receptors which trigger function upon ligand occupation, the initiation of most effector functions by Fc receptors requires additional receptor crosslinking upon Fc domain-FcR interactions. Engagement of Fc receptors by crosslinking with ligand or mAb initiates a cascade of signalling events, including an increase in $[Ca^{2+}]_i$, formation of inositol phosphates, and protein phosphorylation. Signal transduction by FcR shares many features with those of T and B cell antigen receptors which collectively belong to the family of multi-chain immune recognition receptors (MIRR) and share various signalling subunits[1] (Fig. 3.1).

GENERAL VIEW OF CELL ACTIVATION THROUGH IMMUNORECEPTORS

Many of these signalling subunits contain a common sequence motif in their cytoplasmic tails known as the immunoreceptor tyrosine-based activation motif (ITAM).[2] It consists of six conserved amino acid residues spaced precisely over an amino acid sequence $(D/EX_7D/EX_2YX_2LX_7YX_2L)$[3] (Fig. 3.2). Its predominant occurrence in receptor chains that were involved in signal transduction and its location in cytoplasmic tails suggest a role in interaction with cytoplasmic signalling effectors.

The cytoplasmic tail of TCR-ζ contains three copies of the ITAM motif (Fig. 3.2). It was shown capable of signal transduction when placed in the context of the "inert" transmembrane-spanning and extracellular domains of CD8.[4] Studies demonstrated that TCR-ζ and the FcεRIγ chain tails could transduce signals leading to cellular cytotoxicity, production of IL-2, and basophil degranulation.[5] Wegener et al

Cell-Mediated Effects of Immunoglobulins, edited by Wolf Herman Fridman and Catherine Sautès. © 1997 R.G. Landes Company.

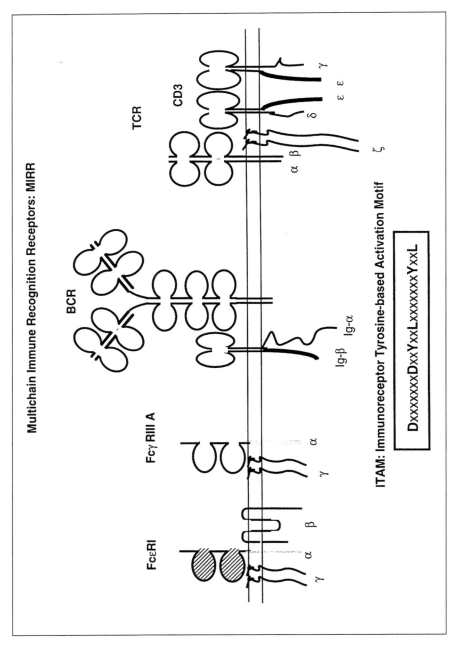

Fig. 3.1. Schematic representation of multichain immune recognition receptors (MIRR). TCR for T cell antigen receptor, BCR for B cell antigen receptor and FcR for the receptors for the Fc portion of immunoglobulins. To simplify, composition of only canonical receptors are presented. Indeed, TCR ζ chain may be substituted by the FcεRI γ chain or by a splicing variant of the ζ chain, the η chain; in addition to Ig-α/Ig-β heterodimer, BCR may be composed of cytoplasmic tail-deleted isoform of Ig-β, named Ig-γ. FcεRI may be expressed on human mast cells without β chain and in contrast, FcγRIIIA may be associated to the FcεRI β chain on murine mast cells or to the TCR ζ chain on human NK cells. In addition, most of the Fc receptors (FcεRI, FcαR, FcγRI, FcγRIIA, FcγRIIB and FcγRIIIA) have been found to be associated to the FcεRI γ chain (see Fig. 3.3). ITAM is defined in Figure 3.2.

also reported that the cytoplasmic tail of CD3ε, can also transduce a signal. The minimal signalling unit in CD3ε was contained in 22 residues defined as CD3ε ITAM.[7] A shorter 18-amino acid ITAM fragment of TCRζ was sufficient to mediate Ca^{2+} mobilization and cytotoxicity.[8] This amino acid fragment represented the ζa ITAM minus the first conserved motif (D/E) residue. Mutations of either of the two tyrosines in the motif abolished signalling.[8] These studies demonstrated that ITAM sequences contain all structural information required for signal transduction leading to cytolysis, Ca^{2+} mobilization, protein tyrosine phosphorylation, and IL-2 production. Other receptor ITAMs or receptor tails containing ITAMs were shown able to transduce signals. These included Ig-α, Ig-β, CD3γ, CD3δ, and FcγRIIA.[2] Therefore, certain individual residues within the motif, in particular the two conserved tyrosines, were of critical importance for this function. ITAMs could thus provide a functional interface with cytoplasmic signalling pathways (Fig. 3.2).

ITAM:	D/E	X	X	X	X	X	X	X	X	D/E	X	X	Y	X	X	L/I	X	X	X	X	-	X	X	X	Y	X	X	L/I
CD3 ζ a	E	T	A	A	N	L	Q	D	P	N	Q	L	Y	N	E	L	N	L	G	R	-	R	E	E	Y	D	V	L
CD3 ζ b	K	Q	Q	R	R	R	N	P	Q	E	G	V	Y	N	A	L	Q	K	D	K	M	A	E	A	Y	S	E	I
CD3 ζ c	E	R	R	R	G	K	G	H	-	D	G	L	Y	Q	G	L	S	T	A	T	-	K	D	T	Y	D	A	L
CD3 γ	D	K	Q	T	-	L	L	Q	N	E	Q	L	Y	Q	P	L	K	D	R	E	-	Y	D	Q	Y	S	H	L
CD3 δ	E	V	Q	A	-	L	L	K	N	E	Q	L	Y	Q	P	L	R	D	R	E	-	D	T	Q	Y	S	R	L
CD3 ε	N	K	E	R	P	P	P	V	P	N	P	D	Y	E	P	I	R	K	G	Q	-	R	D	L	Y	S	G	L
Ig-α	D	M	P	D	-	D	Y	E	D	E	N	L	Y	E	G	L	N	L	D	D	-	C	S	M	Y	E	D	I
Ig-β	D	D	G	K	A	G	M	E	E	D	H	T	Y	E	G	L	N	I	D	Q	-	T	A	T	Y	E	D	I
FcεRI-γ	A	A	I	A	S	R	E	K	A	D	A	V	Y	T	G	L	N	T	R	N	-	Q	E	T	Y	E	T	L
FcεRI-β	E	L	E	S	K	K	V	P	D	D	R	L	Y	E	E	L	N	H	V	Y	-	S	P	I	Y	S	E	L

Fig. 3.2. How immunoreceptor tyrosine-based activation motif (ITAM) has been defined. The basic observation has been made by M. Reth in 1989 who aligned the amino-acid sequences of cytoplasmic tails from various subunits associated to antigen receptors. The corresponding sequences are shown (CD3γ, CD3δ, CD3ε, CD3ζabc, Ig-α, Ig-β, FcεRI γ chain, FcεRI β chain). He defined the consensus sequence (DxxxxxxDxxYxxLxxxxxxYxxL) as antigen receptors homology 1 domain or ARH1. During the next years, several laboratories demonstrated that this domain is the key structure in the triggering of cell activation though antigen and Fc receptors by tyrosine kinases. The domain was thus functionally defined as immunoreceptor tyrosine-based activation motif (ITAM).

An extensive literature has demonstrated the tyrosine phosphorylation of TCR, BCR, and FcR subunits during signal transduction. The mutational analyses of chimeric receptors suggested that this phosphorylation occurs in the ITAM and indicated that phosphorylation of both conserved ITAM tyrosines may be necessary for signal transduction.

For the receptors containing multiple ITAMs, a general scheme of the signal transduction cascade involved in cell activation may be proposed (Fig. 3.3). The first event following crosslinking, is the binding of Src family PTK to ITAM, resulting in activation of Src family PKTs and ITAM tyrosine phosphorylation.[9,10] This step is followed by the

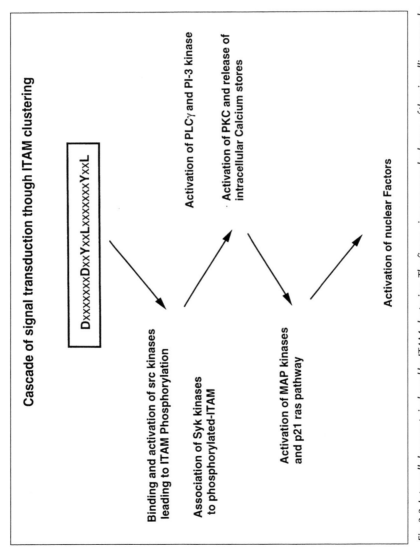

Fig. 3.3. Intracellular events induced by ITAM clustering. The figure gives a general scheme of the signalling cascade induced by antigen receptors through ITAMs.

translocation/association of Syk family members (i.e. p72[Syk] or ZAP70) with phosphorylated ITAM.[11-13] Downstream in the signalling cascade phospholipase C (PLC)γ1, PLCγ2, and phosphatidylinositol-3 (PI-3) kinase also become activated.[14,-17] This results in the conversion of phosphatidyl-inositol 4,5-biphosphate (PIP2) to diacylglycerol and phosphatidyl-inositol 1,4,5- triphosphate (IP3), leading to activation of protein kinase C (PKC) and Ca^{2+} release from intracellular stores. PKC catalyses serine/threonine phosphorylation which may modulate interactions with other tyrosine kinases/phosphatases. Receptor stimulation, furthermore, leads to tyrosine phosphorylation of other proteins such as MAP kinase, the GTPase activating protein for p21[ras] (p62[Gap]), and the product of the proto-oncogene p95[Vav]. The function of the latter protein is not clear, but may be related to cell growth regulation. Signal transduction processes through tyrosine phosphorylation require the activity of both tyrosine kinases and tyrosine phosphatases (PTPases) such as CD45 or the SHP (also known as PTP1C) which are transmembrane PTPase expressed by most of hematopoietic cells.[18,19]

A general view of signal transduction via FcR could be that the ability of FcR to trigger cell activation should be determined by the presence of ITAM in the molecular complex constituting FcR. Except for the low affinity receptor for IgE or CD23, which is a type II membrane receptor and has structural homology with lectin molecules,[20] most of the FcR are constituted of a ligand binding subunit member of the Ig superfamily which is associated to various molecules bearing ITAMs.[2] One exception is an isoform of human FcR for IgG (FcγIIA) which has a nontypical ITAM in the cytoplasmic tail of the ligand binding subunit.[21] Three distinct associated chains, with one or several ITAMs, are a part of molecular complexes which constitute FcRs (Fig. 3.4). The γ chain is a disulphide-linked homodimer which was first described as a component of the high affinity receptor for IgE[22] then found in receptors for IgG[23,24,25] and for IgA.[26] The ζ chain, initially described as a disulphide-linked homodimer associated to the CD3 complex, was also found in human low affinity receptors for IgG.[27,28] Lastly, the β chain is a tetraspan membrane molecule specifically found in the high affinity receptor for IgE and a low affinity receptor for IgG[22,29] in mast cells. In addition to the high conservation of the components of FcR, they have specific functions at the surface of immunocompetent cells. For example, the receptors for the Fc portion of IgG (FcγR) mediate lymphokine secretion by natural killer (NK) cells, antibody-dependent cell cytotoxicity (ADCC) by macrophages,[30] degranulation of mast cells[31] and inhibition of membrane Ig-induced B cell activation.[32] The wide variety of biological responses elicited by engagement of FcR can be envisioned as cell type-specific and as a FcR-specific response to the interaction with immune complexes. To clarify the involvement of each FcR in cell activation, they must be analyzed independently.

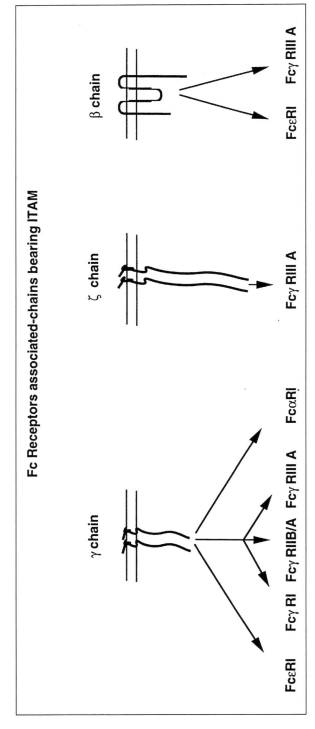

Fig. 3.4. Multiple combinations of chains bearing ITAMs with FcR. The figure depicts the associations of γ, ζ and β chains with the different ligand binding subunits of Fc receptors.

A MODEL OF FcR WITH SPECIFIC FUNCTION:
THE Fc RECEPTOR FOR IgE (FcεRI)

The high affinity receptor for IgE (FcεRI) is composed of three subunits: an α chain, a β chain and a pair of γ chains.[21] Chemical crosslinking and biosynthetic studies established this stoichiometry for the rat receptor and in view of the sequence similarity between the rat, mouse and human subunits, the tetrameric structure was proposed as the generic one.[33] The sequences of each subunit provided information about those portions of each subunit that are on the outside, inside, and in the middle of the plasma membrane.[22]

Signalling through FcεRI is initiated by specific binding of multivalent antigen to receptor-bound IgE which results in receptor aggregation. The earliest intracellular events are receptor tyrosine phosphorylation, which is detectable within 5 seconds after receptor aggregation and activation of the lyn tyrosine kinase.[34,35] After 15-30s, syk and btk tyrosine kinases become phosphorylated and intracellular calcium is mobilized.[36-38] Release of secretory granule contents and synthesis and release of arachidonic acid metabolites begin within 1-5 minutes, while later events such as cytokine synthesis begin within 30-60 minutes.[39,40] Molecular interactions through FcεRI which connect kinase activation to intermediary and late events determine the mechanisms of mast cell activation.

When placed in the context of chimeric receptors, individual ITAM motifs have been found to have cell activation properties equivalent to those of their intact parent receptors.[41] However, the activity of FcεRI cannot be reduced to the cumulative effects of the four ITAMs which constitute the receptor. Association of Lyn with FcεRI supported receptor clustering-dependent β and γ chain phosphorylation and recruitment and phosphorylation of Syk. Nonaggregated FcεRI receptors are associated with kinase that can phosphorylate exogenous peptide substrates but does not phosphorylate receptor subunits. After aggregation, receptor-associated kinase activity increases linearly with time, and receptor subunits are phosphorylated. This phosphorylation quickly reaches a plateau, which presumably reflects transphosphorylation of only adjacent receptors. When aggregated receptors are disaggregated by addition of free hapten, receptor complexes retain heightened kinase activity toward peptide substrates but no longer phosphorylate receptor components. Upon FcεRI aggregation, receptor-associated kinase phosphorylates endogenous substrates, presumably receptor ITAMs, by transphophorylation. As a consequence, the kinase is further activated. Data from the lab of J.-P. Kinet and others have suggested that the FcεRI α and β subunits play different roles in signalling and have led to a proposed model of FcεRI activation in which β and γ act cooperatively (Fig. 3.5). The tyrosine kinase lyn is linked to β chain under resting conditions. After receptor engagement with antigen, it phosphorylates β and γ. The phosphorylated γ chain binds syk which becomes phosphorylated and activated by lyn.[42-44] The complex constituted by γ subunit and syk is therefore a "cell activation module" (Fig. 3.5). This model is supported by data which show that mutation of the canonical tyrosines in the γ ITAM eliminates the ability of FcεRI to activate cells.[45]

The role of β chain in intracellular signalling is important to understand how FcεRI signals in various human cells since the human α chain may be expressed at the cell surface as both $αγ_2$ and $αβγ_2$ receptors.[46] Therefore, the β chain may play a role in determining how particular cells respond in vivo to IgE FcεRI-dependent signalling. The contribution of the β chain to cell activation is difficult to establish in rat and in mouse since, unlike human $αγ_2$ complexes, rat and murine $αγ_2$ complexes are not transported to the surface.[22] A murine model generated by transgenic expression of human α chain in wild type and β knockout backgrounds would permit the study of the effects of β signal amplification on IgE/FcεRI-mediated signalling in a native environment.

Fig. 3.5. Signal transduction through FcεRI. One of the Fc receptors expressed by mast cells (FcεR I) is a very interesting model to analyze the cooperation between different ITAMs having specific intracellular effectors (lyn for the β chain and syk for the γ chain), the activation of which leads to mast cell degranulation.

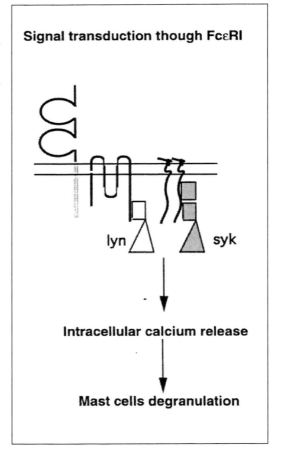

FcεRI signalling is controlled by negative regulation of antigen receptor-mediated activation of tyrosine kinases. This is due to its ability to be rapidly engaged by multi-hapten antigen and disengaged by addition of an excess of monovalent hapten after multi-hapten antigen stimulation. The phosphatases involved in this process are unknown. In contrast, the phosphatase PTP1C has been recently involved in the negative regulation of intracellular signalling though BCR.[19] Indeed, co-crosslinking of FcγRIIB1 with the B cell receptor has been shown to inhibit both BCR-mediated calcium mobilization and IL-2 synthesis.[47] This function is attributable to a 13 amino acid motif within the FcγRIIB1 tail.[47] Signalling through FcεRI and the BCR are both known to require ITAM phosphorylation and result in the activation of syk. In addition, FcγRIIB1 is present on mast cells,[48] suggesting that FcγRIIB1 may negatively regulate signalling through FcεRI. The recruitment of a tyrosine phosphatase into the microenvironment of FcεRI ITAMs could be a potent regulator of mast cell activation though this receptor.

A MODEL OF MULTIFUNCTIONAL FcR: THE LOW AFFINITY Fc RECEPTORS FOR IgG (FcγRII AND FcγRIII)

The low-affinity receptors for the Fc portion of IgG (FcγRII and FcγRIII) bind aggregated or complexed IgG but not monomeric IgG. In contrast, high affinity receptors (FcγRI) can also bind monomeric IgG and are specifically expressed on monocytes and macrophages.[21] The structure of these receptors has been extensively analyzed in both mice and humans. Two isoforms of type II FcγR, FcγRIIB1 and FcγRIIB2, are generated by alternative splicing of sequences coding for a portion of the cytoplasmic domain[49] (Fig. 3.6). The FcγRIIB1 isoform is mainly expressed on B and T lymphocytes,[49-51] whereas FcγRIIB2 is found on macrophages.[52] Type III FcγR (FcγRIII) is a multimeric receptor composed of three different subunits:[49] an α chain and a homodimer of γ chains or ζ chains (Fig. 3.6). The α chain is the ligand binding subunit, and is 95% homologous to FcγRII in the extracellular domains and less than 20% homologous to FcγRII in the transmembrane and the cytoplasmic regions[49] (Fig. 3.6). The γ chain was initially described as a component of the high affinity receptor for the Fc portion of IgE (FcεRI).[22] The homodimeric ζ chain has been initially described as a component of the CD3 complex of the T lymphocyte antigen receptor (TCR).[53] Natural killer (NK) cells express only the FcγRIII,[51,54] in contrast to macrophages, mast cells and certain pre-B cell lines that co-express this receptor with FcγRIIB2 or FcγRIIB1.[48,49,53] In addition to the forms described above, two other FcγR species exist in humans. Neutrophils express a glycosyl-phosphatidylinositol (GPI)-linked form of FcγRIII (FcγRIIIB) which is not associated with the γ chain.[55] Human macrophages and monocytes express a third form of FcγRII, named FcγRIIA, composed of extracellular and transmembrane domains homologous to the murine FcγRII, and a completely different cytoplasmic tail containing a nontypical ITAM.[21]

In contrast to human FcγR, the extracellular regions of murine FcγRII and FcγRIII bind the same ligands and are recognized by the same monoclonal antibody.[56,57] Thus, murine low-affinity FcγR has been an attractive experimental model to investigate the role of different cytoplasmic domains in cell activation after crosslinking of nearly identical extracellular domains by the same ligand.

In mice, the role of FcγRII and FcγRIII in cell activation has been investigated by expressing these receptors in a mast cell line and B lymphoma cells. When expressed by transfection in rat basophilic leukemia cells, FcγRIII, but not FcγRII, induces release of serotonin, arachidonic acid metabolites and hexaminidase but also secretion of tumor necrosis factor (TNF).[58,59] When expressed in FcγR-negative B lymphoma cells, FcγRIII, but not FcγRII, triggers a rise in cytoplasmic Ca^{2+}, PTK activation and cytokine secretion.[60] A mutant receptor lacking the cytoplasmic domain of the ligand-binding α chain, efficiently activated B lymphoma cells. In contrast, the same cytoplasmic-deleted α chain expressed in the absence of γ chain did not transduce activation signals.[60] Signal transduction through FcγRIII is therefore independent of the cytoplasmic region of the ligand-binding chain but requires the associated γ chains.[58-61] In agreement with these observations, the cytoplasmic domain of the associated γ chain, fused to the extracellular and transmembrane domains of other receptors, confers to the chimeric proteins the ability to trigger cell activation. This was found in a variety of systems, including mast cells, T and B lymphocytes[4,5,58-60] Mutational analysis identified two critical tyrosine residues in the cytoplasmic region of the murine γ chain that are part of structural motif involved in cell activation via murine FcγRIII[60] (Fig. 3.6).

FcγRIIB are not directly involved in cell activation but seems mainly able to negatively regulate intracellular signalling passing though ITAM and their cytoplasmic ligands. FcγRIIB contain in their cytoplasmic tail a sequence, named ITIM for immunoreceptor tyrosine-based inhibition motif[2] (see Fig. 3.7, and chapter 4). This motif has been shown critical for down-modulation of BCR-induced $[Ca^{2+}]_i$ increase and cytokine production.[47] Co-crosslinking of FcγRIIB1 with the BCR results in phosphorylation of FcγRIIB1-ITIM leading to recruitment of the phosphatase PTP-1C via its SH2 domain.[19] A critical role for this phosphatase activity in regulation of Ig production was, further, demonstrated in vivo in deficient moth-eaten mice.

SPECIFIC FUNCTIONS OF SEVERAL Fc RECEPTORS

FcγRIIIA

FcγR are involved in cell activation in a wide variety of cell-types. In NK cells, macrophages, mast cells, and polymorphonuclear cells, binding of immune complexes or anti-FcγR antibodies lead to activation of the cells.[30,62] FcγRIIIA has so far been involved in cell activation. Human NK cells have been first used to investigate the role of FcγRIIIA in cell activation. Crosslinking of this receptor by a monoclonal antibody triggers calcium release from intracellular stores, an increase of inositol triphosphate (IP3) and synthesis of lymphokines.[63,64] Human FcγRIIIA expressed by T lymphoma cell,[61] triggers a rise in intracellular Ca²⁺ and activation of protein tyrosine kinases (PTK). The transmembrane form of FcγRIII (FcγRIIIA), but not the GPI-anchored form (FcγRIIIB), mediates cell activation,[61] suggesting that the association to γ chains is required. Interestingly, engagement of human FcγRIIIA in non-lymphoid cells did not lead to any of these cell activation events.[61,65] The cell type but also the components associated with FcγRIIIA may be an important way to determine the effects of cell stimulation by immune complexes. On NK cells, FcγRIIIA can either associate with γ2 or ζ2 homodimers, or γ-ζ heterodimers. Co-transfection studies in COS showed FcγRIIIA/γ to be ≈6 times more efficient than FcγRIIIA/ζ.[66] This quantitative difference between both signalling subunits was shown to reside in the two internal xx amino acids within the YxxL sequences of ITAM.[67]

On NK cells, the crosslinking of FcRγIIIA induces tyrosine phosphorylation of ζ chain,[68] and association of the Src-family PTK p56[Lck]

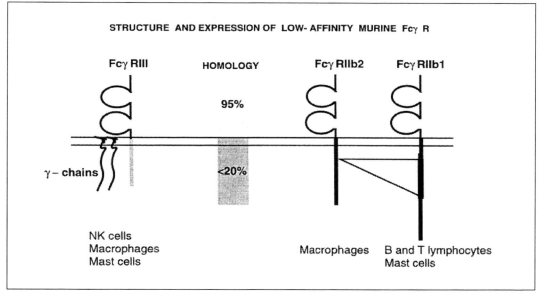

Fig. 3.6. Structure of murine low affinity receptors for IgG. Mouse FcγRII/III were an important model of Fc receptors to investigate how their intracytoplasmic sequences may determine specific functions. Indeed, mouse low affinity FcγR have very homologous extracellular domains (95%) but differ for their cytoplasmic tails. Furthermore, they have specific tissue distribution.

which triggers tyrosine phosphorylation of PI-3 kinase, PLCγ1 and PLCγ2.[15,16,68-71] Both PTK ZAP70 and p72[Syk][16] and nuclear factors, NFATp and NFATc[72] were implicated in FcγRIIIA-induced activation of cytokine genes in human NK cells. On macrophages, the FcγRIIIA signalling cascade is less well documented than for NK cells. However, FcγRIIIA crosslinking leads also to rapid tyrosine phosphorylation of p72Syk[11,13] and association of p95[Vav] and p62[Gap].[73]

FcγRIIA

This receptor contains a unique nontypical ITAM in its cytoplasmic tail. Mutational analyses showed the tyrosines within this motif to be critical for increases in $[Ca^{2+}]_i$, tyrosine phosphorylation and

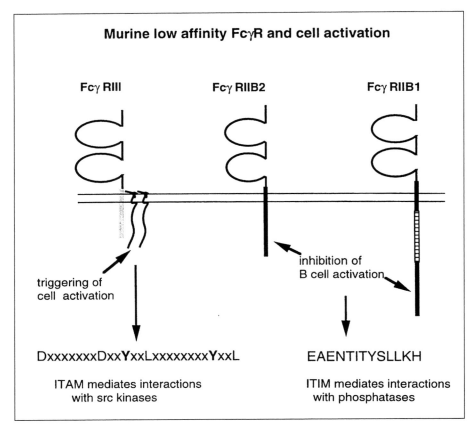

Murine low affinity FcγR and cell activation

Fcγ RIII   Fcγ RIIB2   Fcγ RIIB1

triggering of
cell activation

inhibition of
B cell activation

DxxxxxxxDxxYxxLxxxxxxxxxYxxL   EAENTITYSLLKH

ITAM mediates interactions   ITIM mediates interactions
with src kinases   with phosphatases

Fig. 3.7. Sequences of Fc receptors involved in cell activation. The analysis of murine low affinity receptors for IgG allowed to demonstrate that the opposite effects of these receptors on cell activation (induction cell activation for FcγRIII, inhibition of B cell activation for FcγRII) is depending on their capacity to interact with tyrosine kinases, for FcγRIII, (via the ITAM of the γ chain) or to interact with the tyrosine phospatase PTP1C, for FcγRII, (via the ITIM of their cytoplasmic tail).

phagocytosis, but not for immune complex internalization.[65,74] In addition, FcγRIIA can physically associate with γ chain subunits in cultured monocytes and alveolar macrophages.[25] A functional association between FcγRIIA and the γ chain was shown in a transfection model.[74,75] This study documents qualitative differences between FcγRIIA ITAM and γ chain ITAM in triggering biological functions. Cytokine can be triggered much more effectively by the γ chain ITAM than via FcγRIIA-ITAM.[74]

One of the earliest biochemical events observed following FcγRIIA engagement is tyrosine phosphorylation of multiple cellular proteins, including the receptor itself.[65,76] In addition, transfection of different FcγRIIA-mutants established a critical role for tyrosine(s) located inside and outside FcγRIIA-ITAM.[74,77] Different phosphorylated molecules may associate to FcγRIIA upon crosslinking. These include the Src family kinases p50[Hck], p56[Lyn] and p58[Fgr] and the PTK p72[Syk]. FcγRIIA crosslinking leads also to the phosphorylation of PLCγ1, PCLγ2, p95[Vav], p62[Gap], and Shc in transfected P388.D1 cells.[11,12,14,80]

FcγRI

The two Src-family kinases, p59[Hck] and p56[Lyn], are involved in FcγRI signalling in monocytes, and in vitro kinase assays show an increase in specific p59[Hck] activity upon crosslinking.[81,82] Tyrosine phosphorylation of p72[Syk], PLCγ1, PLCγ2, p95[Vav], and p62[Gap] in activated FcγRI complexes was documented in monocytes.[11,12,83,84] FcγRI crosslinking, furthermore, triggers a mobility shift in MAP kinase.[82] Studies on monocytic HL60 cells demonstrate the *cbl* protooncogene product p[120c-cbl] to be tyrosine phosphorylated upon FcγRI engagement, and to directly associate to p56[Lyn].[85] A critical role was shown for CD45 in FcγRI signalling.[12] In addition, serine/threonine phosphorylation of γ chain has been implicated to play a role in FcγRI signal transduction.[83]

FcαR

These molecules have been detected on most populations of phagocytic cells in blood and mucosal tissues. Engagement of these molecules can trigger phagocytosis, degranulation, oxidative burst, inflammatory mediator release, and antibody-dependent cellular cytotoxicity. FcαR on monocytes/macrophages and neutrophils has been defined as a 55-75 kDa glycoprotein, whereas the eosinophil FcαR is more heavily glycosylated (70-100 kDa). Both types of myeloid receptors are recognized by the CD89 mAb. They bind both IgA1 and IgA2 via their Fc regions. The cDNA encoding the myeloid FcαR has been characterized and was found to encode a 30 kDa peptide, with two extracellular Ig-like domains, a hydrophobic transmembrane region and a cytoplasmic tail devoid of recognized signalling motifs.

Following transfection, FcαR was expressed at the surface of IIA1.6 cells by itself, but lacked signalling capacity.[26] By co-transfection experiments, γ chain was shown able to mediate signal transduction via FcαR. Co-expression of FcαR and FcR γ chain in IIA1.6 B cells conferred signalling capacity to FcαR.[26] In the U937 cell lines, FcαR was reported to associate with γ chain which become phosphorylated on tyrosine residues following FcαR crosslinking.[86] γ chain seems to be critical for FcαR-mediated transmembrane signal transduction but cell type-specific analysis of its signalling capabilities remain to be performed.

A NON-TYPICAL Fc RECEPTOR: THE LOW-AFFINITY RECEPTOR OF IgE: (CD23)

The low-affinity receptor for IgE (FcεRII or CD23) is the only FcR which does not belong to the immunoglobulin superfamily. This molecule has been discovered independently either as an IgE receptor on human lymphoblastoid B cells,[87] as a cell surface marker specifically expressed on Epstein-Barr virus-transformed B cells (RBVCS)[88] or as a B cell activation antigen (Blast 2).[89] FcεRII displays a significant homology with a large family of animal lectins, suggesting the existence of natural ligands other than IgE. The CD23 antigen may be found on a very large variety of cells including B cells, T cells, monocytes, eosinophils, platelets, follicular dendritic cells, Langerhans cells, a subset of thymic epithelial cells, and EBV-containing nasopharyngeal carcinoma cells.[20,90] FcεRII may be viewed not only as a low-affinity IgE receptor expressed on the surface of several cell types. It is also the membrane precursor of a soluble lymphokine with pleiotropic activities.

The CD23 gene codes for two transcripts (named A and B) differing in their 5' untranslated sequences and in their intracytoplasmic region.[91] The expression of type A CD23 is restricted to B cells, whereas the type B isoform is found in all the other types of cells capable of expressing CD23. Low-affinity receptor for IgE is a type II membrane protein with a short N-terminal intracytoplasmic tail (23 residues), a single transmembrane domain (20 residues) and a large C-terminal extracellular region (277 residues). The ability of CD23 to transduce a signal to B cells,[92] together with the very short length of the intracytoplasmic domain of the molecule, suggest that it may be associated with other molecule(s) involved in signal transduction.

Although all the well-characterized activities of FcεRII/CD23 described to date are IgE-dependent, it is most likely that this molecule has other functions that are thought to be mediated by its lectin-like domain. The IgE-dependent functions of CD23 vary according to the cell type on which it is expressed. On inflammatory cells (monocytes/macrophages, eosinophils and platelets) CD23 was shown to be involved in IgE-dependent cytotoxicity against parasites.[93] Other IgE-dependent functions of CD23 on monocytes include the release of IL-1 and TNF-α[94] and the generation of superoxides.[95] CD23 is also capable of delivering a growth signal to mouse B cells.[96,97] CD23 may regulate B cell proliferation however, depending upon the experimental conditions, engagement of CD23 may either enhance or inhibit the progression of B cells from G1 to S phase of the cell cycle.

Supporting the role of CD23 in B cell activation is its ability to signal via the phosphoinositide pathway.[92] Interestingly, engagement of CD23 on the surface of monocytic cells fails to trigger phosphatidyl inositol hydrolysis and the associated intracellular [Ca^{2+}] increase. The engagement of membrane CD23 in IL-4-activated monocytes stimulates a delayed increase in cyclic AMP (cAMP) and elicits the production of various inflammatory products, such as TNF-α and lipid mediators. The CD23-driven production of these mediators can be suppressed by inhibitors of the NOS pathway and activation of the NO pathway through CD23 has been observed in human monocytes, keratinocytes and eosinophils, and in rat macrophages.

CD23 engagement induces a growth arrest in mature monocyte/macrophage-like cells through a mechanism involving the NO pathway. In promonocytic cells, ligation of CD23 elicits a rapid stimulation of the L-arginine-dependent accumulation of cGMP and cAMP, followed by activation of the transcription factor NK-κB, induction of proto-oncogenes c-fos, c-jun, jun-B and c-fms, production of TNF-α and, finally, terminal differentiation into monocytes/macrophages.

CD23 promotes transcription of the gene encoding iNOS in human monocytes/macrophages resulting in production of catalytically active 135 kDa iNOS protein, as revealed by the conversion of ^{14}C L-arginine to ^{14}C L-citrulline and the generation of nitrites. However, the precise mechanism of CD23-linked intracellular events that lead to induction of the gene encoding iNOS remains to be clarified.[98]

ACKNOWLEDGMENTS

C. Bonnerot thanks A. Leray for her excellent secretarial assistance during the preparation of this manuscript.

REFERENCES

1. Keegan AD, Paul WE. Multichain immune recognition receptors: similarities in structure and signaling pathways. Immunol Today 1992; 13:63-68.
2. Cambier JC. New nomenclature for the Reth motif (or ARH1/TAM/ARAM/YXXL). Immunol Today 1995; 16:110-114.
3. Reth M. Antigen receptor tail clue. Nature 1989; 338:383.
4. Irving BA, Weiss A. The cytoplasmic domain of the T cell receptor ζ chain is sufficient to couple to receptor-associated signal transduction pathways. Cell 1991; 64:891-909.
5. Letourneur F, Klausner RD. T cell and basophil activation through the cytoplasmic tail of T cell receptor ζ family proteins. Proc Natl Acad Sci USA 1991; 88:8905-89010.

6. Wegener AMK, Letourneur F, Hoeveler A et al. The T cell receptor/CD3 complex is composed of at least two autonomous transduction modules. Cell 1992; 68:83-95.

7. Letourneur F, Klausner RD. Activation of T cells by a tyrosine kinase activation. Domain in the cytoplasmic tail of CD3ε. Science 1992; 255:79-84.

8. Romeo C, Amiot M, Seed B. Sequence requirements for induction of cytolysis by the T cell antigen/Fc receptor zeta chain. Cell 1992; 68:889-897.

9. Saouaf SJ, Mahajan S, Rowley RB et al. Temporal differences in the activation of three classes of non-transmembrane protein tyrosine kinases following B cell antigen receptor surface engagement. Proc Natl Acad Sci USA 1994; 91:9525-28.

10. Burkhardt AL, Stealey B, Rowley RB et al. Temporal regulation of non-transmembrane protein tyrosine kinase enzyme activity following T cell antigen receptor engagement. J Biol Chem 1994: 269;123642-47.

11. Agarwal A, Salem P, Robbins KC. Involvement of p72^{syk}, a protein-tyrosine kinase, in Fcγ receptor signaling. J Biol Chem 1993; 268:15900-5.

12. Kiener PA, Rankin BM, Burkhardt AL et al. Cross-linking of Fcγ receptor I (FcγRI) and receptor II (FcγRII) on monocytic cells activates a signal transduction pathway common to both Fc receptors that involves the stimulation of p72 Syk protein tyrosine kinase. J Biol Chem 1993; 268:24442-48.

13. Greenberg S, Chang P, Silverstein SC. Tyrosine phosphorylation of the γ subunit of Fcγ receptors, p72^{syk}, and paxillin during Fc receptor-mediated phagocytosis in macrophages. J Biol Chem 1994; 269:3897-903.

14. Shen Z, Lin CT, Unkeless JC. Correlations among tyrosine phosphorylation of Shc, p72^{syk}, PLC-γ1, and $[Ca^{2+}]_i$ flux in FcγRIIa signalling. J Immunol. 1994; 152:3017-23.

15. Azzoni L, Kamoun M, Salcedo TW et al. Stimulation of FcγRIIIA results in phospholipase C-γ1 tyrosine phosphorylation and p56lck activation. J Exp Med 1992; 176:1745-50.

16. Ting AT, Karnitz LM, Schoon RA et al. Fcγ receptor activation induces the tyrosine phosphorylation of both phospholipase C (PLC)-γA and PLC-γ2 in natural killer cells. J Exp Med 1992; 176:1751-58.

17. Kanakaraj P, Duckworth B, Azzoni L et al. Phosphatidylinositol-3 kinase activation induced upon FcγRIIIa-ligand interaction. J Exp Med 1994; 179:551-58.

18. Okumura M, Thomas ML. Regulation of immune function by protein tyrosine phosphatases. Curr Opin Immunol 1995; 7:312-319.

19. D'Ambrosio D, Hippen KL, Minskoff SA et al. Recruitment and activation of PTP1C in negative regulation of antigen receptor signaling by FcγRIIb1. Science 1995; 269:293-97.

20. Delespesse G, Suter U, Mossalayi et al. Expression, structure and function of the CD23 antigen. Adv Immunol 1991; 49:149.

21. Ravetch JV, Kinet JP. Fc receptors. Ann Rev Immunol 1991; 9:457-492.

22. Blank U, Ra C, Miller L et al. Complete structure and expression in transfected cells of high affinity IgE receptor. Nature 1989; 337:187-9.

23. Ernst LK, Duchemin A-M, Anderson CL. Association of the high-affinity receptor for IgG (FcγRI) with the γ subunit of the IgE receptor. Proc Natl Acad Sci USA 1993; 90:6023-27.

24. Scholl PR, Geha RS. Physical association between the high-affinity IgG receptor (FcγRI) and the γ subunit of the high-affinity IgE receptor (FcεRIγ). Proc Natl Acad Sci USA 1993; 90:8847-50.

25. Masuda M, Roos D. Association of all three types of FcγR (CD64, CD32, CD16) with a γ chain homodimer in cultured human monocytes. J Immunol 1993; 151:6382-88.

26. Morton HC, Van den Herik-Oudijk IE, Vossebeld P et al. Functional association between the human myeloid immunoglobulin A Fc receptor (CD89) and FcR γ chain. J Biol Chem 1995; 270:29781-87.

27. Kurosaki T, Gander I, Ravetch JV. A subunit common to an IgG Fc receptor and the T cell receptor mediates assembly through different interactions. Proc Natl Acad Sci USA 1991; 88:3837-41.

28. Anderson P, Calugiuri M, O'Brien C et al. FcR receptor type III (CD16) is included in the ζ NK receptor complex expressed by human natural killer cells. Proc Natl Acad Sci USA 1990; 87:2274-81.

29. Kurosaki T, Gander I, Wirthmueller U et al. The β subunit of the FcεRI is associated with the FcγRIII on mast cells. J Exp Med 1992; 175:447-460.

30. Segal DM, Snider DP. Targeting and activation of cytotoxic lymphocytes. Chem Immunol 1989; 47:179-207.

31. Daeron M, Prouvost-Danon A, Voisin GA. Mast cell membrane antigens and Fc receptors in anaphylaxis. II. Functionally distinct receptors of IgG and IgE on mouse mast cells. Cell Imunol 1980; 40:178-189.

32. Phillips NE, Parker DC. Fc-dependent inhibition of mouse B cell activation by whole anti-μ antibodies. J Immunol 1983; 130:602-609.

33. Metzger H, Alcaraz G, Hohman R et al. The receptor with high affinity for immunoglobulin E. Ann Rev Immunol 1986; 4:419-450.

34. Paolini R, Jouvin MH, Kinet JP. Phosphorylation and dephosphorylation of the high-affinity receptor for immunoglobulin E immediately after receptor engagement and disengagement. Nature 1991; 353:855-858.

35. Eiseman E, Bolen JB. Engagement of the high-affinity IgE receptor activates src protein-related tyrosine kinases. Nature 1992; 355:78-80.

36. Benhamou M, Ryba NJ, Kihara H et al. Protein-tyrosine kinase p72syk in high affinity IgE receptor signaling. Identification as a component of pp72 and association with the receptor γ chain after receptor aggregation. J Biol Chem 1993; 268:23318-24.

37. Kawakami Y, Yao L; Miura T et al. Tyrosine phosphorylation and activation of Bruton tyrosine kinase upon FceRI cross-linking. Mol Cell Biol 1994; 14:5108-13.

38. Maeyama K, Hohman RJ, Metzger H et al. Quantitative relationships between aggregation of IgE receptors, generation of intracellular signals, and histamine secretion in rat basophilic leukemia (2H3) cells. Enhanced responses with heavy water. J Biol Chem 1986; 261:2583-92.

39. Garcia-Gil M, Siraganian RP. Source of the arachidonic acid released on stimulation of rat basophilic leukemia cells. J Immunol 1986; 136:3825-28.

40. Grabbe J, Welker P, Moller A et al. Comparative cytokine release from human monocytes, monocyte-derived immature mast cells, and a human mast cell line (HMC-1). J Invest Dermatol 1994; 103:504-8.

41. Eiseman E, Bolen JB. Signal transduction by the cytoplasmic domains of FcεRI-γ and TCR-ζ in rat basophilic leukemia cells. J Biol Chem 1992; 267:21027-32.

43. Jouvin MH, Adamczewski M, Numerof, R et al. Differential control of the tyrosine kinases Lyn and Syk by the two signaling chains of the high affinity immunoglobulin E receptor. J Biol Chem 1994; 269:5918-25.

44. Alber G, Miller L, Jelsema CL et al. Structure-function relationships in the mast cell high affinity receptor for IgE. Role of the cytoplasmic domains and of the β subunit. J Biol Chem 1991; 266:22613-20.

42. Kihara H, Siraganian RP. Src homology 2 domains of Syk and Lyn bind to tyrosine-phosphorylated subunits of the high affinity IgE receptor. J Biol Chem 1994; 269:22427-32.

45. Paolini R, Renard V, Vivier E et al. Different roles for the FcεRI γ chain as a function of the receptor context. J Exp Med 1995; 181:247-55.

46. Miller L, Blank U, Metzger H et al. Expression of high-affinity binding of human immunoglobulin E by transfected cells. Science 1995; 244:334-7.

47. Amigorena S, Bonnerot C, Drake J et al. Cytoplasmic domain heterogeneity and functions of IgG Fc receptors in B lymphocytes. Science 1992; 256:1808-12.

48. Benhamou M, Bonnerot C, Fridman WH et al. Molecular heterogeneity of murine mast cell Fcγ receptors. J Immunol 1990; 144:3071-80.

49. Ravetch JV, Luster AD, Weinshank R et al. Structural heterogeneity and functional domains of murine immunoglobulin G Fc receptors. Science 1986; 234:718-21.

50. Bonnerot C, Daeron M, Varin N et al. Methylation if the 5' region of the murine β FcγR gene regulates the expression of Fcγ receptor II. J Immunol 1988; 141:1026-34.

51. Bonnerot C, Amigorena S, Fridman WH et al. Unmethylation of specific sites in the 5' region is critical for the expression of murine αFcγR gene. J Immunol 1990; 144:323-8.

52. Lewis VA, Koch T, Plutner H et al. A complementary DNA clone for a macrophage-lymphocyte Fc receptor. Nature 1986; 324:372-5.

53. Ra C, Jouvin MHE, Blank U et al. A macrophage Fcγ receptor and the mast cell receptor for IgE share an identical subunit. Nature 1989; 341:752-4.

53. Weissman AM, Baniyash M, Hou D et al. Molecular cloning of the ζ chain of the T cell antigen receptor. Science 1988; 239:1018-22.

54. Perussia B, Tutt MM, Qiao WQ et al. Murine natural killer cells express functional Fcγ receptor II encoded by the FcγRα gene. J Exp Med 1989; 170:73-85.

55. Ravetch JV, Perussia B. Alternative membrane forms of FcγRIII (CD16) on human NK cells and neutrophils: cell-type specific expression of two genes which differ in single nucleotide substitutions. J Exp Med 1989; 170:481-93.

56. Unkeless JC. Characterization of monoclonal antibodiy directed against mouse macrophage and lymphocyte Fc receptors. J Exp Med 1979; 150:580-596.

57. Weinshank RL, Luster AD, Ravetch JV. Function and regulation of a murine macrophage-specific IgG Fc receptor, FcγR-α. J Exp Med 1988; 167:1909-21.

58. Daeron M, Bonnerot C, Latour S et al. Murine recombinant FcγRIII, but not FcγRII, trigger serotonin release in rat basophilic leukemia cells. J Immunol 1992; 149:1365-73.

59. Latour S, Bonnerot C, Fridman WH et al. Induction of tumor necrosis factor-α production by mast cells via Fcγ R. Role of the FcγRIII γ subunit. J Immunol 1992; 149:2155-62.

60. Bonnerot C, Amigorena S, Choquet D et al. Role of associated γ chain in tyrosine kinase activation via murine FcγRIII. EMBO J 1992; 11:2747-57.

61. Wirthmueller U, Kurosaki T, Murakami MS et al. Signal transduction by FcγRIII (CD16) is mediated through the γ chain. J Exp Med 1992; 175:1381-90.

62. Anderson CL, Looney RJ. Review: human leukocyte IgG Fc receptors. Immunol Today 1986; 7:264-6.

63. Anegon I, Cuturi MC, Trinchieri G et al. Interaction of Fc receptor (CD16) ligands induces transcription of interleukin 2 receptor (CD25) and lymphokine genes and expression of their products in human natural killer cells. J Exp Med 1998; 167:452-64.

64. Carsatella MA, Anegon I, Cuturi M et al. FcγR (CD16) interaction with ligand induces Ca^{2+} mobilization and phosphainositide turnover in human natural killer cells. Role of Ca^{2+} in FcγR (CD16)-induced transcription and expression of lymphokine genes. J Exp Med 1989; 169:549-65.

65. Huang MM, Indik ZK, Brass LF et al. Activation of FcγRII induces tyrosine phosphorylation of multiple proteins including FcγRII. J Biol Chem 1992; 267:5467-73.

65. Odin JA, Edberg JC, Painter CJ et al. Regulation of phagocytosis and [Ca^{2+}]$_i$ flux by distinct regions of an Fc receptor. Science 1991; 254:1785-88.

66. Park J-G, Isaacs RE, Chien P et al. In the absence of other Fc receptors, FcγRIIIa transmits a phagocytic signal which requires the cytoplasmic domain of its γ chain subunit. J Clin Invest. 1993; 92:1976-83.

67. Park J-G, Schreiber AD. Determinants of the phagocytic signal mediated by the type IIIa Fcγ receptor, FcγRIIIa: sequence requirements and interaction with protein-tyrosine kinases. Proc Natl Acad Sci USA 1995; 92:7381-85.

68. Vivier E, Morin P, O'Brien C et al. Tyrosine phosphorylation of the FcγRIII (CD16): ζ complex in human natural killer cells. J Immunol 1991; 146:206-10.

69. O'Shea JJ, Weissman AM, Kennedy ICS et al. Engagement of the natural killer cell IgG Fc receptor results in tyrosine phosphorylation of the ζ chain. Proc Natl Acad Sci USA 1991; 88:350-4.

70. Pignata, C, Prasad KVS, Robertson MJ et al. FcgRIIIA-mediated signalling involves src-family lck in human natural killer cells. J Immunol 1993; 151:6794-6800.

71. Salcedo TW, Kurosaki T, Kanakaraj P et al. Physical and functional association of p56lck with FcγRIIIA (CD16) in natural killer cells. J Exp Med 1993; 177:1475-83.

72. Aramburu J, Azzoni, L, Rao A et al. Activation and expression of the nuclear factors of activated T cells, NFATp and NFATc, in human natural killer cells: regulation upon CD16 ligand binding. J Exp Med 1995; 182:801-10.

73. Darby C, Geahlen RL, Schreiber AD. Stimulation of macrophage FcγRIIIA activates the receptor-associated protein tyrosine kinase Syk and induces phosphorylation of multiple proteins including p95Vav and p62/GAP-associated protein. J Immunol 1994; 152:5429-37.

74. Van den Herik-Oudijk IE, Ter Bekke MWH, Tempelman MJ et al. Functional differences between two Fc receptor ITAM signaling motifs. Blood 1995; 86:3302-7.

75. Van den Herik-Oudijk IE, Capel PJA, van der Brugge et al. Identification of signaling motifs with human FcγRIIa and FcγRIIb isoforms. Blood 1995; 85:2202-11.

76. Ghazizadeh S, Bolen JB, Fleit HB. Physical and functional association of Src-related protein tyrosine kinases with FcγRII in monocytic THP-1 cells. J Biol Chem 1994; 269:8878-84.

77. Mitchell MA, Huang M, Chien P et al. Substitutions and deletions in the cytoplasmic domain of the phagocytic receptor FcγRIIa: effect on receptor tyrosine phosphorylation and phagocytosis. Blood 1994; 84:1753-59.

78. Hamada F, Aoki M, Akiyama et al. Association of immunoglobulin G Fc receptor II with Src-like protein-tyrosine Fgr neutrophils. Proc Natl Acad Sci USA 1993; 90:6305-09.

79. Hunter S, Huang MM, Indik ZK et al. FcγRIIa-mediated phagocytosis and receptor phosphorylation in cells deficient in the protein tyrosine kinase Src. Exp Hematol 1993; 21:1492-97.

80. Liao F, Shin HS, Rhee SG. Tyrosine phosphorylation of phospholipase C-γ1 induced by cross-linking of the high-affinity or low-affinity Fc receptor for IgG in U937 cells. Proc Natl Acad Sci USA 1992; 89:3659-63.

81. Wang AVT, Scholl PR, Geha RS. Physical and functional association of the high affinity immunoglobulin G receptor (FcγRI) with the kinases Hck and Lyn. J Exp Med 1994; 180:1165-70.

82. Durden DL, Min Kim H, Calore B et al. The FcγRI receptor signals through the activation of Hck and MAP kinase. J Immunol 1995; 154:4039-47.

83. Durden DL, Rosen H, Cooper JA. Serine/threonine phosphorylation of the γ-subunit after activation of the high-affinity Fc receptor for immunoglobulin G. Biochem J 1994; 299:569-77.

84. Durden DL, Liu YB. Protein-tyrosine kinase p72^{syk} in FcγRI receptor signaling. Blood 1994; 84:2102-08.

85. Marcilla A, Rivero-Lezcano OM, Agarwal A et al. Identification of the major tyrosine kinase substrate in signaling complexes formed after engagement of Fcγ receptors. J Biol Chem 1995; 270:9115-20.

86. Pfefferkorn LC, Yeaman GR. Association of IgA-Fc receptors (FcαR) with FcεRIγ2 subunits in U937 cells. Aggregation induces the tyrosine phosphorylation of γ2. J Immunol 1994; 153:3228-35.

87. Gonzalez-Molina A, Spiegelberg HL. Binding of IgE myeloma proteins to human cultured lymphoblastoid cells. J Immunol 1976; 117:1838-45.

89. Thorley-Lawson DA, Nadler LM, Bhan AK et al. BLAST-2 (EBVCS), an early cell surface marker of human B cell activation, is superinduced by Epstein-Barr virus. J Immunol 1985; 134:3007-16.

88. Kintner C, Sugden B. Identification of antigenic determinants unique to the surfaces of cells transformed by Epstein-Barr virus. Nature 1981; 294:458-62.

90. Delespesse G, Hofstetter H, Sarfati M et al. Human FcεRII. Molecular, biological and clinical aspects. Chem Immunol. 1989; 47:79-102.

91. Yokota A, Kikutani H, Tanaka T et al. Two species of human Fcε receptor II (FcεRII/CD23): tissue-specific and IL-4 specific regulation of gene expression. Cell 1988; 55:611-25.

92. Kolb JP, Renard D, Dugas B et al. Monoclonal anti-CD23 antibodies induce a rise in intracellular calcium and polyphosphoinositide hydrolysis in human activated B cells. Involvement of a GP protein. J Immunol. 1990; 145:429-36.

93. Capron A, Dessaint JP, Capron M et al. From parasites to allergy: the second receptor for IgE (FcεR2). Immunol Today 1986; 7:15-18.

94. Borish L, Mascali JJ, Rosenwasser LJ. IgE-dependent cytokine production by human peripheral blood mononuclear phagocytes. J Immunol 1991; 146:63-71.

95. Kikutani H, Yokota A, Uchibayashi N et al. Structure and function of Fcε receptor II (FcεRII/CD23): a point of contact between the effector phase of allergy and B cell differentiation. Ciba Found Symp 1989; 147:23-42.

96. Campbell KA, Lees A, Finkelman FD et al. Crosslinking FcεRII and surface IgD enhances anti-IgD mediated B cell activation. FASEB J 1991; 5:1384-9.

97. Waldschmidt TJ, Tygrett L. Crosslinking surface Ig and the low affinity IgE Fc receptor induces B cells to enter cell cycle. FASEB J 1991; 5:1389.

98. Dugas B, Mossalayi MD, Damais C et al. Nitric oxide production by human monocytes: evidence for a role of CD23. Immunol Today 1995; 16:574-80.

CHAPTER 4

FcγR as Negative Coreceptors

Marc Daëron

THE IMMUNORECEPTOR FAMILY

Receptors involved in the recognition of antigen by cells of the immune system are of three types: B cell receptors (BCR), T cell receptors (TCR) and Fc receptors (FcR). In spite of having different structures, of being expressed with different tissue distributions, of recognizing antigen under distinct modalities and of triggering different cellular responses, BCR, TCR and FcR were recently understood as structurally and functionally related members of the same family, referred to as the immunoreceptor family.

BCR are expressed exclusively by B lymphocytes. They recognize native antigens as they come into contact with the organism. They are composed of an antigen recognition subunit, consisting of a membrane immunoglobulin molecule, and of heterodimeric Igα-Igβ signal transduction subunits.[1]

TCR are expressed exclusively by T lymphocytes. They recognize peptides derived from the intracellular degradation of protein antigens inserted in the peptide groove of major histocompatibility complex (MHC) molecules expressed at the surface of antigen-presenting cells. They are composed of a peptide + MHC recognition subunit, consisting of an α-β or γ-δ heterodimer, and of three signal transduction subunits: two heterodimers, CD3ε-δ and CD3ε-γ, and one homodimer, TCRζ.[2]

FcR are expressed by many cells of the lymphoid and myeloid lineages. They do not recognize antigens but the Fc portion of antibodies. Nevertheless, antibody-FcR complexes, which form when antibodies

Cell-Mediated Effects of Immunoglobulins, edited by Wolf Herman Fridman and Catherine Sautès. © 1997 R.G. Landes Company.

bind to FcR, function as membrane receptors for antigen with an ad-
justable specificity. Upon binding to FcR, antibodies thus provide antigen
specificity to a great variety of effector cells devoid of antigen recognition structures. Indeed, when aggregated by antibodies and multivalent antigens, FcR can trigger the biological activities of such cells,
enabling them to respond specifically to antigens. FcR are composed
of an immunoglobulin-binding α subunit associated or not to signal
transduction subunit(s). Most FcR are multichain receptors the α chain
of which is associated with a homodimeric FcRγ subunit.[3] FcRγ associates with the α subunits of high-affinity receptors for IgE (FcεRI),
IgA (FcαRI) or IgG (FcγRI), and to the α subunit of low-affinity receptors for IgG (FcγRIIIA).

   FcRγ and TCRζ are functionally and structurally related.[4] They
are both hardly exposed to the outside—the extracellular domain of
FcRγ has only five residues, that of TCRζ nine residues—and they
possess a transmembrane domain having a charged residue, involved
in the association with the other subunits. They can each associate
either to FcR or to TCR and, in some cases, they can form γ-ζ
heterodimers: in human NK cells, the α subunit of FcγRIIIA is associated with TCRζ instead of FcRγ; in some T cells, TCR are expressed
in association with FcRγ instead of TCRζ. In mast cells, FcR which
possess FcRγ also associate with a single-chain FcRβ subunit with four
transmembrane domains[5] and whose amino- and carboxy-terminal ends
are both intracytoplasmic.[6]

   Homologies between BCR, TCR and FcR were further empha-
sized when Michael Reth noticed the presence of a common motif, in
the intracytoplasmic domains of Igα, Igβ, CD3ε, CD3δ, CD3γ, TCRζ,
FcRγ and FcRβ.[7] Such a motif was also found in the intracytoplasmic
domain of a single-chain low-affinity receptor for IgG unique to humans and referred to as FcγRIIA. This motif, composed of a twice-
repeated YxxL sequence flanking 7-12 variable residues, was soon understood to account for the cell-triggering properties of receptors
possessing these subunits (Fig. 4.1). Initially described under several
names, activation motifs expressed in BCR, TCR and FcR were recently renamed and collectively designated as immunoreceptor tyrosine-
based activation motifs (ITAMs).[8] Upon receptor aggregation, ITAMs
become rapidly tyrosine-phosphorylated by one or several protein tyrosine
kinases of the *src* family. Phospho-ITAMs can then recruit cytoplasmic
molecules which have SH2 domains. They thus provide docking sites
for cytoplasmic protein tyrosine kinases of the *syk* family—ZAP70 in
T cells, ZAP70 and *syk* in NK cells, *syk* in other cells—whose two
SH2 domains bind each to one phosphorylated YxxL motif of ITAMs.[9]
Phospho-ITAM-bound *syk* or ZAP70 become phosphorylated by *src*
kinases and activated. They then phosphorylate other substrates among
which is phospholipase Cγ which initiates the turnover of phosphati-
dylinositides, itself resulting in calcium mobilization and, ultimately,

in cell activation. Intriguingly, BCR, TCR and FcR combine variable associations of different ITAMs: human FcγRIIA have only one, FcR associated to FcRγ possess two (one in each monomer of FcRγ), FcR associated to FcRγ and FcRβ three, BCR probably four (one in each Igα and Igβ molecule), as one assumes that two Igα-Igβ heterodimers associate to one membrane immunoglobulin, and TCR 10 (one ITAM in each chain of the two CD3 heterodimers and three in each chain of the TCRζ homodimer). The functional significance of such an abundance of ITAMs with the same consensus di-Y-L structure intermingled

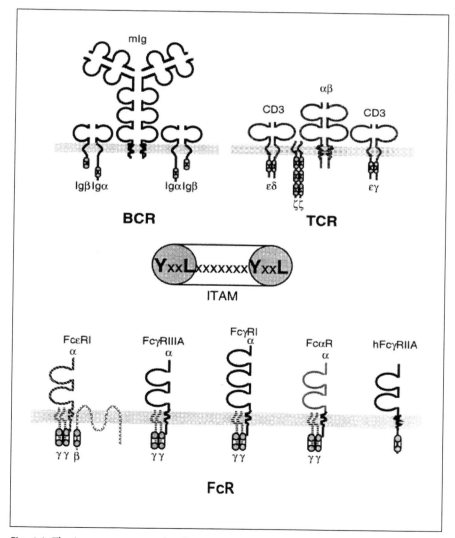

Fig. 4.1. The immunoreceptor family.

with fixed numbers of varying "x" residues remains unclear inasmuch as, upon aggregation, monomeric chimeric molecules having a single ITAM were found to trigger cell activation.

If FcR which possess ITAMs can trigger cell activation, not all FcR possess ITAMs. FcγRIIB are such receptors and they cannot activate cells.[10,11] FcγRIIB are single-chain low-affinity IgG receptors which are highly homologous in mice and humans.[12-17] In both species, they are encoded by a single gene in which a separate exon encodes the transmembrane domain and three other exons the intracytoplasmic domain.[18] This structural peculiarity—the α subunits of all FcR associated with FcRγ are encoded by genes in which a single exon encodes the transmembrane and the intracytoplasmic domains—enables several isoforms to be generated by alternative splicing of corresponding sequences. Two such human FcγRIIB isoforms were described: FcγRIIB1, which retain sequences encoded by all four exons, and FcγRIIB2, which lack sequences encoded by the first intracytoplasmic exon. Murine FcγRIIB exist as three membrane isoforms and one soluble isoform. As in humans, FcγRIIB1 possess sequences encoded by all four exons and FcγRIIB2 lack sequences encoded by the first IC exon. The FcγRIIB1-specific insertion encoded by the first IC exon is longer, however, in the murine receptor (47 amino acids) than in the human receptor (19 amino acids) because the murine first intracytoplasmic exon has 84 extra 3' nucleotides which are absent in the corresponding human exon. An additional murine isoform, resulting from the use of a cryptic splice donor site, located where the human exon stops in the first intracytoplasmic murine exon, was recently described by us.[19] Rather than FcγRIIB1, it is the actual murine homologue of human FcγRIIB1 and, for this reason, we named it FcγRIIB1'. Murine FcγRIIB3 lack sequences encoded by the transmembrane and the first intracytoplasmic exons, and they are released as soluble molecules by macrophages. FcγRIIB are widely expressed by cells of hematopoietic origin. FcγRIIB1 are preferentially expressed by cells of the lymphoid lineage, FcγRIIB2 by cells of the myeloid lineage and murine FcγRIIB1' by cells of both lineages.

Although unable to trigger cell activation, membrane FcγRIIB are not devoid of biological properties. When aggregated by multivalent ligands, they are involved in capping, endocytosis and phagocytosis. Sequences responsible for these biological activities were recently mapped in the intracytoplasmic domain of murine and human FcγRIIB. A tyrosine-containing 13-amino acid sequence, encoded by the third intracytoplasmic exon, was shown to determine the ability of both murine and human FcγRIIB2 to mediate the endocytosis of soluble immune complexes via clathrin-coated pits.[20] Another tyrosine-containing sequence, more carboxy-teminal, accounts for the ability of murine FcγRIIB2 to trigger the phagocytosis of particulate immune complexes.[21] Human FcγRIIB2, in which the carboxy-terminal tyrosine involved in phagocytosis by murine FcγRIIB2 is not conserved, do not trigger phagocytosis.[22] Interestingly, sequences encoded by the first intra-cytoplasmic exon were found to inhibit the internalization properties of carboxy-terminal sequences.[23] They probably account for the inability of murine FcγRIIB1 and FcγRIIB1' to mediate endocytosis and phagocytosis and for the inability of human FcγRIIB1 to mediate endocytosis. Since these sequences are reduced to the 19 amino-terminal residues encoded by the 5' end of first intracytoplasmic exon in human FcγRIIB1 and murine FcγRIIB1', sequences which inhibit internalization could be mapped within these residues. Finally, the same 19 amino acids encoded by the first intracytoplasmic exon were found to be necessary and sufficient for capping,[24] a property of FcγRIIB1 and FcγRIIB1', when aggregated at 37°C.[19] In both humans and mice, FcγRIIB therefore appear as a subfamily of receptors, with a wide tissue distribution, unable to activate cells, and whose biological properties depend on the various combinations of effector and regulatory sequences, in the intracytoplasmic domain which characterizes each isoform (Fig. 4.2).

But FcγRIIB exert another major biological function when they are coaggregated to other immunoreceptors, instead of being simply aggregated to themselves. By contrast with antigen receptors which are clonally expressed on lymphocytes—all BCR molecules on a given B cell possess the same antigen-binding structure and all TCR molecules on a given T cell possess the same peptide+MHC-binding structure—FcR

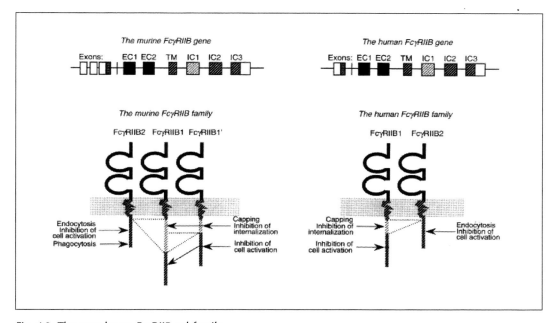

Fig. 4.2. The membrane FcγRIIB subfamily.

are indeed not clonally distributed, and a single cell usually expresses more than one type of FcR. On many cells, FcγRIIB are coexpressed with immunoreceptors which can trigger cell activation. FcγRIIB are expressed together with BCR on B cells,[25] together with TCR on activated T cells,[26,27] together with FcεRI and FcγRIIIA on mast cells[28] and Langerhans cells,[29] together with FcγRI and FcγRIIIA on macrophages.[30] This enables FcγRIIB to become coaggregated with adjacent immunoreceptors. Under these conditions, FcγRIIB were found to regulate negatively cell activation triggered by ITAM-containing immunoreceptors. This property is the subject of the present review.

FcγRIIB INHIBITS BCR-MEDIATED B CELL ACTIVATION

It has been known for 30 years that IgG antibodies to a given antigen can inhibit antibody responses to that antigen. Passively administered IgG antibodies were first shown to inhibit in vivo primary responses[31] and IgG antibodies were found soon to be effective on primary in vitro responses.[32] These early experiments (reviewed in ref. 33) prompted immunologists to adopt the endocrinology-derived concept of negative feedback regulation for antibody production. IgG antibodies did not seem to inhibit antibody responses by masking epitopes, rendering antigen unable to be recognized by immunocompetent cells since, while inhibiting antibody production, they did not prevent priming for a secondary response.[34] IgG antibodies could thus induce a state of active tolerance resulting from the deviation of immunological responses from the synthesis of effector molecules towards the constitution of an immunological memory. The concept was extended to other antibody-dependent immunoregulatory phenomena such as allotype suppression, induced in rabbits and mice by an injection of anti-allotype antibodies,[35] and idiotype suppression, induced by injecting mice with anti-idiotype antibodies of various origins.[36] Antigen-specific, allotype-specific and idiotype-specific suppression of antibody production all focused attention on membrane immunoglobulins that had previously been understood as being the antigen receptors of B cells.

A decisive advance in unraveling mechanisms underlying these immunoregulatory effects was made when antibody-induced inhibition was demonstrated to depend on the Fc portion of the suppressive antibodies. Chan and Sinclair showed first that in order to inhibit primary anti-sheep red blood cell (SRBC) antibody response in vivo, anti-SRBC IgG have to possess an intact Fc portion.[37] The same was found to apply to antigen-specific adoptive tolerance induced by the transfer of spleen cells preincubated with IgG antibodies[38] and to primary in vitro antibody responses.[39] Likewise, allotype suppression and idiotype suppression were not induced by F(ab')2 fragments of anti-allotype[35] or anti-idiotype[40] antibodies respectively. One step further, it was reported that antigen-specific inhibition induced by incubating mouse spleen cells with rabbit anti-SRBC IgG antibodies, which was lost when the Fc protion of antibodies was removed, could be restored if anti-SRBC F(ab')2 were aggregated with intact IgG anti-rabbit immunoglobulins.[41] Finally, anti-IgM IgG antibodies were found to be unable to activate B cells, whereas F(ab')2 fragments of the same antibodies triggered polyclonal B cell activation.[42] These observations altogether supported the concept, first proposed by Sinclair as the *tripartite model inactivation* of B cells, according to which the aggregation of membrane immunoglobulins on B cells, by multivalent antigens, anti-allotype, anti-idiotype or anti-isotype F(ab')2, would lead to B cell activation, whereas the coaggregation of membrane immunoglobulins to FcγR by corresponding intact antibodies or IgG immune complexes, on the same cell, would induce a state of B cell inactivation[37] (Fig. 4.3). Such a coaggregation, indeed, not only failed to activate B cells, it induced a long lasting inhibition. This conclusion was based on the finding that the incubation of mouse spleen cells with intact rabbit IgG antibodies to murine IgM inhibited the proliferative responses of spleen cells to all B cell mitogens.[43] Under the same conditions F(ab')2 fragments were 20 times less active.

Indirect evidence for a role of B cell FcγR in antibody-mediated inhibition was provided first by Stockinger and Lemmel who showed that neither the depletion of T cells nor that of macrophages prevented inhibition, whereas that of FcγR-positive cells did.[41] Direct evidence came from the use of the rat anti-mouse FcγR monoclonal antibody

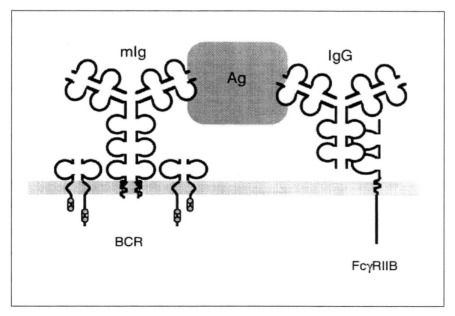

Fig. 4.3. Sinclair's tripartite model of B cell inactivation.

2.4G2 which was then described.[44] Indeed, intact IgG anti-mouse immunoglobulin antibodies could stimulate B cells if FcγR were made inaccessible to the Fc portion of antibodies by preincubating cells with 2.4G2.[45] After cDNAs encoding murine FcγR were cloned in 1986, and the distinction of three low-affinity FcγR was made,[12-14] mouse B cells were found to express only FcγRIIB1.[25] Definitive evidence for the role of this FcγR isoform in inhibition was obtained by reconstitution experiments in IIA1.6 cells, a FcγR-negative variant of the murine lymphoma B cells A20/2J[46] which has a deletion in the FcγRIIB gene. IIA1.6 cells have a functional BCR, made of membrane IgG2a, whose aggregation either by F(ab')2 fragments of rabbit anti-mouse IgG (RAM), or by intact RAM IgG trigger the tyrosine phosphorylation of intracellular substrates, an increase of the concentration of intracellular Ca^{2+} and the secretion of IL-2. IIA1.6 cells were stably transfected with cDNA encoding mouse FcγRIIB1, FcγRIIB2 or FcγRIIB whose intracytoplasmic domain had been deleted, and stimulated with either RAM F(ab')2 or RAM IgG. RAM F(ab')2 triggered the secretion of IL-2 in all three transfectants whereas RAM IgG, which stimulated tail-less FcγRIIB-expressing transfectants, failed to stimulate cells expressing FcγRIIB1 or FcγRIIB2.[24]

These results prompted other groups to investigate whether human FcγRIIB could exert the same inhibition. IIA1.6 cells were stably transfected with human FcγRIIB1, FcγRIIB2 or tail-less FcγRIIB. As for murine receptors, the coaggregation of BCR to human FcγRIIB1 and FcγRIIB2 inhibited cell activation whereas that of BCR to tail-less FcγRIIB did not.[47] Finally, we found recently that, when expressed in IIA1.6 cells, the new murine FcγRIIB isoform, FcγRIIB1', which has the same 19-amino acid insertion encoded by the first intracytoplasmic exon as human FcγRIIB1, is as efficient as the two other FcγRIIB isoforms to inhibit B cell activation.[19]

These data, altogether, accredited Sinclair's original concept that FcγR expressed on B cells function as regulatory molecules enabling IgG antibodies, once generated in response to an antigen stimulation, to regulate specifically their own production in the presence of antigen. Inhibition could be assigned to FcγRIIB1, which are constitutively expressed by B cells, and which would inhibit BCR-induced cell activation through a process requiring intracytoplasmic sequences of the receptor shared by murine and human FcγRIIB1 and FcγRIIB2. That both FcγRIIB isoforms could inhibit B cell activation equally well was somewhat surprising for a regulatory process restricted so far to B cells which do not express FcγRIIB2.

FcγRIIB INHIBITS FcR-MEDIATED MAST CELL ACTIVATION

If, however, one understands how FcεRI and BCR as immunoreceptors share similar ITAMs capable of triggering cell activation via similar intracellular biochemical mechanisms, one may speculate that the regulatory properties of FcγRIIB might not be restricted to B cells.

We and others reported previously that mouse mast cells coexpress FcεRI, FcγRIIIA and FcγRIIB.[28,48] FcγRIIIA and FcγRIIB bind the same subclasses of IgG, and they are both recognized by the rat mAb anti-murine low-affinity FcγR 2.4G2. This made mouse mast cells inappropriate to investigate the possible regulatory properties of murine FcγRIIB on mast cell activation by FcεRI. We therefore constructed a transfectant model in the rat mast cell line RBL-2H3.[49] These cells express functional FcεRI whose aggregation triggers the release of pre-formed granule mediators, such as serotonin, the synthesis of newly-formed lipid mediators, such as leukotrienes, and the secretion of in-flammatory cytokines, such as TNFα. They also express rat FcγRIIIA and FcγRIIB which are not recognized by 2.4G2. RBL cells were sta-bly transfected with cDNAs encoding mouse FcγRIIB1, FcγRIIB2 or FcγRIIB whose intracytoplasmic domain was either deleted or replaced by that of the α subunit of mouse FcγRIIIA which, so far, has not been found to trigger any cellular response. Resulting transfectants released mediators when FcεRI were aggregated in cells sensitized with rat IgE and challenged with F(ab')2 fragments of mouse anti-rat (MAR) im-munoglobulins. They failed to release mediators when recombinant receptors were aggregated with 2.4G2 F(ab')2 fragments and MAR F(ab')2.[10] Cells were therefore sensitized with rat IgE and incubated with or without increasing concentrations of 2.4G2 F(ab')2 fragments before they were challenged with MAR F(ab')2. Mediator release, trig-gered by FcεRI aggregation in cells not exposed to 2.4G2, was dose-dependently inhibited when FcεRI were coaggregated to FcγRIIB in cells incubated with 2.4G2 F(ab')2. Both FcγRIIB1 and FcγRIIB2 were inhibitory, but not FcγRIIB without intracytoplasmic domain or with that of FcγRIIIA.[50]

Similar stable transfectants were constructed with human FcγRIIB cDNAs. They failed to release mediators when recombinant receptors were aggregated with F(ab')2 fragments of the mouse anti-human FcγRIIA/B mAb AT10 and F(ab')2 fragments of goat anti-mouse (GAM) immunoglobulins, but they responded normally to FcεRI ag-gregation when sensitized with mouse IgE and challenged with GAM F(ab')2. Mediator release was dose-dependently inhibited when FcεRI were coaggregated to FcγRIIB in cells incubated with increasing con-centrations of AT10 F(ab')2. As with murine receptors both human FcγRIIB1 and FcγRIIB2 were inhibitory, but not FcγRIIB without the intracytoplasmic domain.[27]

Since human FcγRIIA also possess an ITAM, RBL transfectants expressing human FcγRIIB1 or FcγRIIB1 without intracytoplasmic domain were retransfected with cDNA encoding human FcγRIIA. When incu-bated with F(ab')2 fragments of the FcγRIIA-specific mouse mono-clonal antibody IV.3 and challenged with GAM F(ab')2, both transfectants released serotonin. Serotonin release was dose-dependently inhibited when FcγRIIA were coaggregated with FcγRIIB1 in cells incubated with F(ab')2 fragments of AT10 which recognize both FcγRIIA and FcγRIIB, and challenged with GAM F(ab')2. No inhibition was seen when FcγRIIA were coaggregated under the same conditions with FcγRIIB without intracytoplasmic domain.[27]

These data altogether indicate that murine and human FcγRIIB can both inhibit mast cell activation triggered either by endogenous rat FcεRI or by recombinant human FcγRIIA. In both species, the FcγRIIB1 and the FcγRIIB2 isoforms were equally inhibitory, provided they had an intact intracytoplasmic domain. Conditions required for inhibition were studied in transfectants expressing mouse FcγRIIB. Inhibition was found to require the coaggregation of the two receptors. It affected only FcεRI coaggregated to FcγRIIB and it was reversible upon receptor deaggregation. Inhibition affected serotonin release as well as the secretion of TNFα.[50]

FcγRIIB INHIBITS TCR-MEDIATED T CELL ACTIVATION

Since, as anticipated, the inhibitory properties of FcγRIIB are not restricted to BCR-mediated B cell activation, but also affect FcR-mediated mast cell activation, one can expect that FcγRIIB might inhibit TCR-mediated T cell activation as well. This was investigated using, as a model system, the murine BW5147 thymoma cells whose TCR had been previously reconstituted by the stable transfection of cDNA encoding mouse CD3ε and TCRζ.[51] These cells were stably transfected with cDNA encoding murine FcγRIIB1, FcγRIIB2 or FcγRIIB without intracytoplasmic domain. Resulting transfectants had a functional TCR whose aggregation by immobilized F(ab')2 fragments of anti-CD3ε antibodies, coated to culture wells, triggered cells to secrete IL-2. IL-2 secretion was dose dependently inhibited if BW5147 transfectants that expressed FcγRIIB1 or FcγRIIB2 were incubated in culture wells coated with a constant concentration of anti-CD3ε F(ab')2 and increasing concentrations of 2.4G2 F(ab')2. Under the same conditions, no inhibition was observed in cells expressing FcγRIIB without intracytoplasmic domain.[27] Like BCR-mediated B cell responses and FcR-mediated mast cells responses, cytokine secretion, induced in a T cell line by aggregating TCR was therefore inhibited when TCR was coaggregated to FcγRIIB having an intact intracytoplasmic domain.

FcγRIIB INHIBITS CELL ACTIVATION BY ALL ITAM-BASED IMMUNORECEPTORS

FcεRI, BCR and TCR are multisubunit receptors which possess three, four and ten ITAMs respectively. By contrast, FcγRIIA are single-chain receptors which possess a single ITAM. Cell activation triggered by the four receptors can be inhibited when coaggregating each ITAM-based receptor to FcγRIIB. This suggests that ITAM-possessing sub-units, and possibly ITAMs themselves are sufficient targets for FcγRIIB to inhibit cell activation. This was investigated using chimeric molecules constructed by joining cDNA sequences encoding the ITAM-containing intracytoplasmic domain of an immunoreceptor and cDNA sequences encoding the transmembrane and extracellular domains of other molecules. Thus, chimeras made of the extracellular and trans-membrane domains of human IgM and of the intracytoplasmic domain of murine Igα (IgM-Igα) or Igβ (IgM-Igβ) were expressed in IIA1.6 cells.[52] Likewise, chimeras made of the extracellular and trans-membrane domains of the human IL-2 receptor α chain and of the intracytoplasmic domain of either FcRγ (IL-2Rα-FcRγ) or TCRζ (IL-2Rα-TCRζ) were stably expressed in RBL cells.[27] The aggregation of IgM-Igα or of IgM-Igβ triggered calcium mobilization in B cells, that of IL-2Rα-FcRγ or IL-2Rα-TCRζ triggered serotonin release in mast cells. Cal-cium mobilization was inhibited when IgM-Igα or IgM-Igβ were coaggregated to murine FcγRIIB, expressed in the same B cells,[52] and serotonin release was inhibited when IL-2Rα-FcRγ or IL-2Rα-TCRζ were coaggregated to murine FcγRIIB, expressed in the same mast cells.[27] Because, TCRζ has three ITAMs, we constructed a IL-2Rα-TCRζ chi-mera whose intracytoplasmic domain was restricted to the 42 amino-terminal residues, i.e., contained the first TCRζ ITAM only, and which was stably expressed in RBL cells. When aggregated at the cell sur-face, this chimera triggered the release of serotonin. Serotonin release was inhibited when it was coaggregated to murine FcγRIIB.[27] It fol-lows that chimeric molecules having a single ITAM—from Igα, Igβ, FcRγ or TCRζ—are sufficient targets for enabling FcγRIIB to inhibit B cell or mast cell activation.

This finding further extends FcγRIIB-dependent inhibition to all immunoreceptors associated with ITAM-containing subunits. If FcγRIIB may regulate TCR-dependent cell activation in T cells which express a TCRαβ or γδ in association with TCRζ, they are likely to regulate as well TCR which express the same peptide+MHC recognition subunits in association with FcRγ. Since FcRγ associates with the ligand-bind-ing α subunit of FcεRI, FcγRI, FcγRIIIA and FcαRI, FcγRIIB may regulate various IgE-, IgG- and IgA-induced responses provided cells coexpress FcγRIIB and corresponding receptors. These include, in mice, mast cells, macrophages, monocytes and cells, and in humans, mac-rophages and monocytes, Langerhans cells, polymorphonuclear neu-trophils, basophils and eosinophils. NK cells are a noticeable excep-tion. These cells express FcγRIIIA[53,54] whose α subunit is associated with FcRγ in mice and to TCRζ in humans, and do not coexpress FcγRIIB which might regulate negatively antibody-dependent cell me-diated cytotoxicity (ADCC). Interestingly, however, murine and hu-man NK cells express inhibitory receptors, which recognize MHC class I molecules on target cells and whose coaggregation with FcγRIIIA inhibits ADCC.[55-57] In humans, the same molecules are also expressed by some T cells and they can inhibit TCR-dependent T cell functions.[58] Murine and human inhibitory receptors belong to structurally unrelated families of molecules. They nevertheless possess a motif which resembles the motif which accounts for the inhibitory properties of FcγRIIB.

THE ITIM MOTIF

Pioneer experiments which first identified an inhibitory sequence in murine FcγRIIB were reported by Amigorena et al. They examined the ability of murine FcγRIIB2 bearing deletions of increasing length in their intracytoplasmic domain to inhibit B cell activation in IIA1.6 transfectants. They found that the deletion of the 16 carboxy-terminal residues of the 47-amino acid intracytoplasmic domain of FcγRIIB2 had no effect on inhibition, whereas the deletion of the next 13 residues abolished inhibition.[24] The inhibitory properties of this 13-amino acid sequence was confirmed by Muta et al who showed that murine FcγRIIB having an intracytoplasmic domain whose 16 carboxy-terminal residues were deleted and whose 18 amino-terminal residues were replaced by those of TCRζ retained an ability to inhibit B cell activation.[52] Combined together, these results indicate that a 13-amino acid intracytoplasmic sequence, comprising intracytoplasmic residues 18 to 31 and common to murine FcγRIIB1 and FcγRIIB2, is both necessary and sufficient to inhibit BCR-dependent B cell activation.

In order to determine the sequence which accounts for inhibition of FcR-dependent mast cell activation, the same FcγRIIB2 deletants were expressed in RBL cells and their effects on serotonin release were examined when they were coaggregated to FcεRI. The same deletions which abrogated inhibition in B cells also abrogated inhibition in mast cells.[27] In the 13-amino acid inhibitory sequence, one finds a tyrosine residue (Y26). There is another tyrosine, in the 16 carboxy-terminal residues which are not required for inhibition (Y43). FcγRIIB2 in the intracytoplasmic domain of which either Y26 or Y43 was replaced by a glycine [FcγRIIB2(G26)] or an alanine [FcγRIIB2(A43)] respectively were expressed in RBL cells. When coaggregated to FcεRI, FcγRIIB2(A43) inhibited serotonin release, but not FcγRIIB2(G26).[27]

In order to determine the sequence which accounts for inhibition of TCR-dependent T cell activation, the same FcγRIIB2 mutants were coexpressed in RBL cells, together with IL-2Rα-TCRζ chimeric molecules, and serotonin release was examined when the two molecules were coaggregated at the cell surface. The deletion of the same 13-amino acid sequence, but not that of the 16 carboxy-terminal residues, abolished the inhibition of IL-2Rα-TCRζ-induced mediator release, as well as the point mutation of Y26, but not that of Y43.[27]

Taken together, these data indicate that the same tyrosine-containing 13-amino acid sequence, in the intracytoplasmic domain of

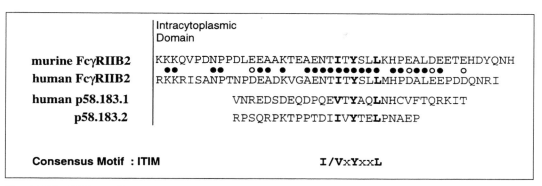

Fig. 4.4. The ITIM motif.

murine FcγRIIB, accounts for inhibition of BCR-dependent B cell activation, of FcR-dependent mast cell activation and of TCR-dependent T cell activation. If one considers that the three immunoreceptors transduce activation signals through common biochemical pathways, this conclusion suggests that inhibition could involve similar mechanisms.

The same approach applied to human FcγRIIB yielded the same results and led to the conclusion that the same 13-amino acid sequence accounts for inhibition of FcεRI-mediated mast cell activation.[27] As stressed earlier, human and murine FcγRIIB are highly homologous. Conserved residues, however, are not evenly distributed along the sequence of the two receptors. As shown by Figure 4.4, they are particularly conserved in an intracytoplasmic stretch which corresponds to the inhibitory sequence. Sequence comparison of the two receptors enables identification of residues which are probably not essential for inhibition. Among conserved residues, one notices the tyrosine whose mutation is sufficient to abrogate the inhibitory properties of the receptor. One also notices that it is followed, at the position Y+3, by a leucine residue. This YslL motif is reminiscent of the double YxxL motif which constitutes ITAMs and determines the cell-triggering ability of immunoreceptors. Although the comparison of human and murine sequences may provide some clues to identify a motif which might account for the inhibitory properties of FcγRIIB, it is not sufficient to define a consensus inhibitory motif. In view of the critical role played by each YxxL of ITAMs in signal transduction by immunoreceptors, the structural analogy with ITAMs, however, was suggestive enough to prompt several groups, including ours, to adopt the acronym ITIM, for *immunoreceptor tyrosine-based inhibition motif*, which designates a still to be defined inhibitory motif, in the FcγRIIB inhibitory sequence. The concept of a YxxL-based inhibitory motif is self evident, and one can readily build up a variety of possible scenarios in which an ITIM could interfere with ITAM-dependent cell signaling. The limited number of inhibitory molecules that can be compared, however, makes the structural definition of ITIM loose, varying from YxxL to 12 of the 13-residue sequence AENTITYSSLKxP (the Y+5 residue not being conserved in humans, it can be omitted). Interestingly, inhibitory receptors expressed by NK cells, in which they inhibit FcγRIIIA-dependent ADCC as well as NK cytotoxicity, and by some T cells, in which they inhibit TCR-dependent cytotoxicity, possess an intracytoplasmic sequence composed of two YxxL motifs separated by 26 residues.[55-57] In view of their inhibitory properties, on the one hand, and of the structural constraints imposed by the interaction of phospho-ITAMs with kinases of the *syk* family on the other hand, this sequence was thought to correspond to two ITIMs, rather than to an ITAM with an usually long intervening sequence between the two YxxL motifs.[59,60] The comparison of sequences flanking these motifs and the FcγRIIB motif, made apparent the conservation of a valine or an isoleucine residue at position Y-2, which is not conserved in ITAMs[59] (Fig. 4.4). If this residue proved to be functionally significant, it should be taken into account in a preliminary structural definition of ITIM which might thus be built on the V/IxYxxL backbone.

BIOCHEMICAL MECHANISMS OF NEGATIVE REGULATION BY FcγRIIB

FcγRIIB-mediated inhibition of B cell, mast cell and T cell activation affects a secretory response. How inhibition affects upstream events leading to these secretory responses has been so far investigated in B cells only.

A critical event for secretion is an increase in the intracellular concentration of Ca^{2+}. This increase results from a rapid and transient efflux of Ca^{2+} from intracellular stores, followed by a sustained influx of extracellular Ca^{2+} across the plasma membrane. The variation with time in the concentration of intracellular Ca^{2+} following the aggregation of BCR or the coaggregation of BCR to FcγRIIB was monitored in LPS-activated murine B cells, A20 lymphoma B cells and IIA1.6 transfectants expressing FcγRIIB1. The results were the same in the three cells. The coaggregation of BCR to FcγRIIB did not affect the initial increase in Ca^{2+} concentration resulting from the mobilization of intracellular Ca^{2+} stores. No subsequent influx of extracellular Ca^{2+}, however, was observed when the two receptors were coaggregated.[61,62] This was found to be an active phenomenon which both prevented the opening of plasma membrane Ca^{2+} channels and closed open Ca^{2+} channels.[62] When studied in IIA1.6 transfectants expressing the various FcγRIIB mutants described earlier, a correlation was found between inhibition of cytokine secretion and inhibition of extracellular Ca^{2+} influx.[52,63]

Mechanisms which account for the different steps of the Ca^{2+} response are far from being fully understood. One generally accepts that the initial rise in intracellular Ca²⁺ results from the generation of inositol phosphates, which results from the activation of phospholipase Cγ (PLCγ).[64,65] Upon activation, PLCγ becomes tyrosine-phosphorylated. The opening of plasma membrane Ca²⁺ channels seems to depend on the increase in the intracellular Ca²⁺ concentration resulting from the release of Ca²⁺ from the endoplasmic reticulum. Its precise mechanism remains unknown. Events involved in this intracellular cascade were examined in A20 cells and in IIA1.6 transfectants. Significant change was observed neither in the total inositol phosphates generated during the first few minutes following the coaggregation of BCR and FcγRIIB, nor in the phosphorylation of PLCγ.[62,52]

Finally, when early phosphorylation events were examined in B cells whose BCR had been coaggregated to FcγRIIB, no obvious inhibition was observed. The phosphorylation of intracellular substrates was not affected in whole cell lysates, and neither the phosphorylation of BCR ITAMs nor that of *syk* seemed to be significantly inhibited.[62,52] These results indicate that inhibition does not result from FcγRIIB preventing BCR aggregation at the cell surface.

That inhibition is apparently not a passive phenomenon was further suggested by the observation that, when coaggregated to BCR, FcγRIIB became tyrosine-phosphorylated.[52] FcγRIIB whose ITIM had the tyrosine mutated failed to become phosphorylated.[63] Since such FcγRIIB mutants are no more capable of inhibiting B cell activation, one may assume that the phosphorylation of the tyrosine residue, in the ITIM, is a critical event during the inhibitory process. The kinase(s) responsible for this phosphorylation has not been identified yet. By analogy with ITAMs, one could anticipate that a phospho-ITIM could be recognized by the SH2 domains present in tyrosine kinases, tyrosine phosphatases, phospholipases and other molecules.[66] To examine this question, synthetic peptides corresponding to nonphosphorylated and phosphorylated ITIM were used as immunoadsorbents to precipitate molecules in cell lysates from A20 B cells. Peptides corresponding to phospho-ITIM, but not peptides corresponding to nonphosphorylated ITIM precipitated three proteins, two of which could be identified as the cytoplasmic phosphatases PTP1C and PTP1D respectively. These protein tyrosine phosphatases were recently renamed as SHP-1 and SHP-2 respectively.[67] SHP-1 and SHP-2 are widely expressed in cells of hematopoetic origin. They have the distinctive characteristic of possessing two tandem SH2 domains[68] reminiscent of the two tandem SH2 domains of the cytoplasmic tyrosine kinases of the *syk* family. Neither syk nor ZAP70, however, bind to the phospho-ITIM peptide. Conversely, FcγRIIB that had been phosphorylated upon coaggregation with BCR were precipitated by GST fusion proteins corresponding to the SH2 domains of SHP-1 incubated with lysates from A20 B cells. Nonphosphorylated FcγRIIB were not precipitated. Finally, small amounts of SHP-1 and of SHP-2 were coprecipitated with FcγRIIB in lysates of A20 B cells whose BCR were coaggregated with FcγRIIB.[63] These data, altogether, led to the conclusion that FcγRIIB-mediated inhibition of BCR-dependent B cell activation may involve the recruitment of cytoplasmic phosphatases having SH2 domains capable of binding to the phospho-ITIM in FcγRIIB. Interestingly, phosphorylated peptides corresponding to the ITIM-like motifs of MHC receptors expressed by NK cells also bound SHP-1 and SHP-2.[59,60] The substrates of the two phosphatases are not known and it would be premature to propose a mechanism by which they could inhibit the generation of B cell responses. One may however hypothesize that, if they are indeed recruited in vivo by the FcγRIIB ITIM, these phosphatases might be functional as d'Ambrosio et al. Found that the phosphatase activity of SHP-1 increased several-fold in the presence of phospho-ITIM, but not in the presence of phospho-ITAM[63] (Fig. 4.5). Surprisingly, neither SHP-1 nor SHP-2 were coprecipitated wiht phosphorylated FcγRIIB in mast cells (M. Daëron and J.C. Caubies, manuscript in preparation), and it was recently reported that the SH2-containing inositol phosphatase SHIP bound to phospho-ITIM.[68b] This suggests

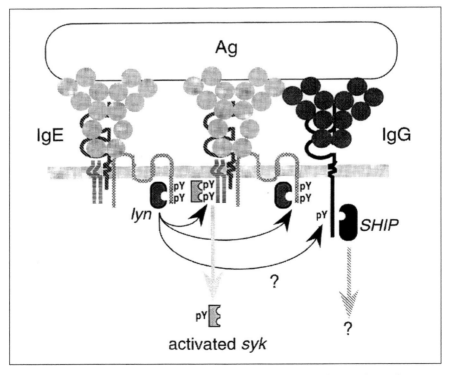

Fig. 4.5. A hypothetical model for FcγRIIB-mediated inhibition of ITAM-dependent mast cell activation.

that several mechanisms might account for FcγRIIb-dependent inhibition of cell activation in different cell types.

BIOLOGICAL SIGNIFICANCE OF NEGATIVE REGULATION BY FcγRIIB

Evidence that FcγRIIB can negatively regulate all ITAM-based immunoreceptors was obtained in rather artefactual experimental models. In many of these models, activating receptors and inhibitory receptors were coaggregated using F(ab')2 fragments of monoclonal antibodies to these receptors rather than natural ligands. One may therefore wonder how biologically significant this conclusion is. The situation is unequal for BCR, TCR and FcR for the following two reasons. First, conditions under which coaggregation can take place are obvious for BCR and FcR since BCR and antibodies can recognize epitopes borne by the same antigen molecule; it is not so for TCR which recognizes a peptide+MHC complex that antibodies do not see. Second, more or less physiological conditions were explored for the three receptors. These were both in vitro and in vivo.

In vitro evidence that IgG immune complexes can inhibit B cell activation has been extensively documented in early works. These were reviewed above.

In vitro evidence that IgG antibodies can negatively regulate IgE-dependent mast cell activation was obtained using bone marrow-derived mast cells (BMMC) obtained by culturing mouse bone marrow cells in the presence of IL-3 for several weeks. Resulting cells consist of a homogeneous population of muscosal-type mast cells which express FcεRI and FcγRIIB1.[28] Upon FcεRI aggregation, but not upon FcγRIIB1 aggregation, these cells release inflammatory mediators. To determine whether FcγRIIB1 could be coaggregated to FcεRI by IgG antibodies and inhibit mediator release under physiological conditions, BMMC were sensitized with anti-ovalbumin IgE antibodies and challenged with DNP-ovalbumin, not complexed or complexed to mouse monoclonal IgG1 anti-DNP antibodies. Non-complexed DNP-ovalbumin triggered IgE-sensitized BMMC cells to release serotonin. Serotonin release was dose-dependently inhibited if DNP-ovalbumin was complexed to increasing concentrations of IgG antibodies.[50] This indicates that, under conditions where no competition for antigen could take place between IgE and IgG antibodies, IgG immune complexes could regulate negatively IgE-dependent serotonin release. It is therefore likely that FcεRI and FcγRIIB1, constitutively expressed by non-transformed mouse mast cells, can be coaggregated by immune complexes the antigen moiety of which binds to the Fab portions of receptor-bound IgE and the antibody moiety of which binds to adjacent FcγRIIB1via their Fc portion. In vitro evidence that constitutive human FcγRIIB can inhibit IgE-dependent basophil activation was also obtained. Human basophils were purified by elutriation from peripheral blood. Human IgE that had bound in vivo to FcεRI were acid-eluted and basophils were passively sensitized with mouse IgE or not. Cells were incubated without or with increasing concentrations of F(ab')2 fragments of the mouse anti-human FcγRII monoclonal antibody AT10, before they were challenged with an excess of F(ab')2 fragments of goat anti-mouse immunoglobulins. Cells sensitized with IgE and not exposed to AT10 F(ab')2 released histamine upon challenge, but not cells incubated with AT10 F(ab')2 and not sensitized with IgE. Histamine release by cells sensitized with IgE was dose-dependently inhibited by AT10 F(ab')2.[27] This indicates that the coaggregation of FcεRI to FcγRII constitutively expressed by human blood basophils inhibits IgE-induced histamine release. It also suggests that the coaggregation of FcγRIIB to FcγRIIA by AT10 F(ab')2 prevented the latter to trigger mediator release. This explains why human basophils are usually unable to respond to IgG immune complexes[69] although they express FcγRIIA. Indeed, human basophils react with the FcγRIIA-specific mouse monoclonal antibody VI.3.[70]

Finally, in vitro evidence that TCR-dependent T cell activation can be negatively regulated by constitutively expressed FcγRIIB1 was so far obtained using T cells lines only. A murine hybridoma T cell, 2B4,[71] and a murine lymphoma T cell, RMA,[72] which express a functional TCR, were found both to react with 2.4G2 in immunofluorescence. They contained only FcγRIIB1 transcripts when analyzed by RT-PCR using FcγRIIB-specific oligonucleotide probes. When plated onto anti-CD3 F(ab')2-coated dishes, the two cells were triggered to secrete IL-2. When plated onto dishes coated with a mixture of anti-CD3 F(ab')2 and of increasing concentrations of 2.4G2 F(ab')2, IL-2 secretion was dose-dependently inhibited.[27] This indicates that the coaggregation of TCR to FcγRIIB1 constitutively expressed by murine T cell lines could inhibit TCR-dependent T cell activation leading to cytokine response.

In vivo evidence of the biological significance of the negative regulation exerted by FcγRIIB is more indirect, especially for FcR and TCR. There are however a number of in vivo observations which have not been given satisfactory explanations and which could be readily accounted for by a negative effect of FcγRIIB.

Historical in vivo experiments reviewed above, which seeded the concept of a negative feedback regulation of antibody production by IgG, amply validate the physiological relevance of the negative effects of FcγRIIB expressed by B cells. Recently published experiments showing that antibody responses are enhanced in FcγRIIB-knock out mice further confirmed the phenomenon.[73] Likewise, *moth-eaten* mice,[74] which bear a genetic defect in the SHP-1 gene and which lack this phosphatase,[75] show polyclonal B cell activation, have a wide array of autoantibodies and possess B cells which can be activated by intact anti-immunoglobulin IgG antibodies.[63] FcγRIIB-dependent inhibition of IgF-induced mediator release was not impaired in BMMC from moth-eaten mice.[68b]

FcγRIIB-knock out mice also provided evidence that the prediction that all ITAM-based immunoreceptors should be susceptible to the inhibitory effects of FcγRIIB might be correct. BMMC were generated from bone marrow cells under conditions which enabled the expression not only of FcγRIIB, but also of FcγRIIIA. BMMC derived under these conditions from normal control mice failed to respond to IgG immune complexes or to receptor aggregation by the anti-FcγRIIB/IIIA monoclonal antibody 2.4G2. BMMC derived from FcγRIIB-deficient mice were readily degranulated in response to the same stimuli. Moreover, IgG-induced passive cutaneous anaphylactic reactions were augmented in FcγRIIB-knock out mice compared to normal mice.[73]

The most suggestive in vivo evidence of the effectiveness of IgG immune complexes in controlling IgE-induced mast cell/basophil activation may be immunotherapy performed in allergic patients, the beneficial effects of which remain unexplained. This treatment consists of injecting increasing doses of allergen to patients, starting with doses that are too low to trigger anaphylactic reactions. As injections are repeated with higher doses, an anti-allergen IgG response develops, and positive responses to treatment may be correlated with the concentration of these antibodies in the serum.[76] The possible role of IgG immune complexes, generated in allergic patients treated by immunotherapy, is supported by reports showing that the injection of preformed immune complexes, made of specific allergen and IgG antibodies to the same allergen, into patients with allergic asthma decreased significantly allergic symptoms.[77,78] Finally, in patients suffering of anaphylactic shock to bee venom, a rapid desensitization protocol is used in which large amounts of antigen are injected within a single day.[79] The treatment may prevent anaphylactic manifestations within hours, i.e., before IgG antibodies could be induced. This suggests that low levels of circulating specific IgG, possibly present in patients before treatment, could form large immune complexes which might coaggregate FcεRI to FcγRIIB and prevent allergen from activating basophils. Such mechanisms, which operate in pathological situations, are likely to operate also under physiological conditions. One may thus speculate that, in normal individuals, mast cells and basophils are constantly stimulated by IgE antibodies generated in response to allergens to which nonallergic individuals are exposed as well as allergic individuals. Cytokines released in IgE-induced reactions may contribute to the regulation of immune responses, and inflammatory mediators could facilitate cytokines to reach their target cells. No allergic symptom appears, however, in normal subjects possibly because allergens become complexed to specific IgG antibodies before they come into contact with mast cells or basophils and keep cell responses under control. Mast cell and basophil FcγRIIB therefore appear as built-in safety devices preventing potentially harmful reactions to develop. Whether such control mechanisms are over-flown in allergic patients is a hypothesis that deserves to be examined.

Evidence supporting the possibility that FcγRIIB might negatively regulate TCR-mediated T cell activation in vivo is scarce. FcγRIIB, expressed on T cells, can conceivably be coaggregated to TCR when T cells recognize peptide+MHC on antigen-presenting cells or target cells in the presence of IgG antibodies directed to epitopes expressed on the same cell. This is susceptible to happen in several situations. The phenomenon of tumor enhancement[80] or the immunological facilitation of normal tissues[81] by passive antibody are two such situations. They require the Fc portion of antibodies,[82] and alloactivated cytotoxic T cells express FcγR.[83] Another situation might be viral infections in which cytotoxic T cells kill infected cells in a MHC-restricted way and can keep infection asymptomatic for long periods. Viruses also induce antibodies against antigens expressed at the surface of infected cells. In some infections, the appearance of symptoms is correlated with an increase in anti-virus antibodies in the serum. Finally, a proportion of serum immunoglobulins were found to be autoantibodies in normal individuals.[84] One may speculate that such autoantibodies, directed to peptide-presenting cells or to invariant epitopes of the TCR itself, could prevent autoreactive T cell clones which escaped from negative selection in the thymus from being harmful. If this were the case, autoimmune manifestations might arise if, for some reason, natural autoantibodies failed to prevent anti-self T cell clones to be activated. The beneficial effects of the intravenous injections of polyclonal immunoglobulins into patients suffering of autoimmune diseases[85,86] might be partly accounted for by such a mechanism.

CONCLUSION

IgG antibodies have long been known as major soluble effector molecules of the immune system. Data discussed in this review make them also major regulators of cell activation during an immune response. IgG antibodies are indeed generated in large quantities in response to most protein antigens and circulate throughout the body in the blood stream. FcγRIIB, on the other hand, are the most ubiquitous of all FcR.[3] The finding that FcγRIIB can regulate negatively all ITAM-based immunoreceptors endows IgG antibodies with a wide spectrum of immunoregulatory properties. IgG antibodies are not only involved in the negative feedback regulation of antibody production, they can also exert regulatory effects at all stages of an immune response. They might possibly affect the onset of an immune response by inhibiting the activation of helper T cells by antigen-presenting cells in the presence of IgG to structures expressed by the latter. They might affect cell-mediated immune responses by inhibiting cytolytic functions of CTL recognizing target cells in the presence of antibodies to molecules expressed by the latter. They might affect inflammatory responses resulting from the interaction of IgE, IgG or IgA with corresponding FcR expressed by mast cells, macrophages or polymorphonuclear cells of the three types. They might affect antibody-dependent cell mediated cytotoxicity exerted by FcγRIIIA-expressing effector cells other than NK cells.

Like most other receptors, FcR have been understood not to be functional when expressed on a plasma membrane. They become functional after they have been aggregated by antibodies and multivalent antigens, and the molecular significance of FcR aggregation has only recently been clarified by the concept of transphosphorylation.[87] Under physiological conditions, however, the aggregation of identical FcR is probably a rare event for the following two reasons. First, FcR of several types, with identical or different isotypic specificities, are coexpressed on most cells of hematopoietic origin. Second, immune complexes have no reason to be made of a single class or subclass of antibodies. As a consequence, when immune complexes interact with FcR on a single cell, they coaggregate different adjacent FcR on the same membrane (Fig. 4.6). The various cell activation-triggering FcR

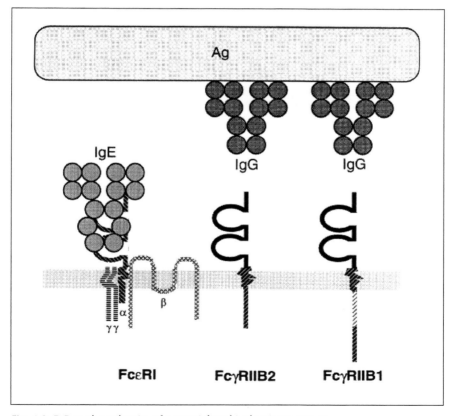

Fig. 4.6. FcR as the subunits of potential multisubunit receptors.

and/or cell activation-regulating FcR that are coexpressed on a single cell thus function as the subunits of multichain receptors which form when they become coaggregated by immune complexes. The qualitative and quantitative composition of multichain receptors of that kind is not predetermined. It depends on the cell type, on cytokines which differentially regulate the expression of the various FcR and on the composition of immune complexes with which they interact. Resulting receptor complexes will transduce signals which, instead of being simply "on" or "off", might establish sophisticated intracellular communication networks through which versatile messages can elaborate. Cell metabolism may be finely tuned by such messages in response to stimuli delivered by environmental factors. Under these conditions, FcR with ITAMs would function as positive coreceptors for each other, FcγRIIB as negative coreceptors for ITAM-based immunoreceptors. It remains to be determined whether their regulatory effect is restricted to receptors with ITAMs.

ACKNOWLEDGMENTS

Experimental data from our laboratory discussed in this review were from the works of Odile Malbec, Sylvain Latour, Sebastian Amigorena, Christian Bonnerot, Eric Espinosa and Patrick Pina. These were supported by the Institut National de la Santé et de la Recherche Médicale, the Institut Curie, Roussel-UCLAF and the Association pour la Recherche sur le cancer

REFERENCES

1. Pleinman C, D'Ambrosio D, Cambier J. The B cell antigen receptor complex: structure and signal transduction. Immunol Today 1994; 15:393-399.
2. Weiss A. T cell antigen receptor signal transduction: a tale of tails and cytoplasmic protein-tyrosine kinases. Cell 1993; 73:209-212.
3. Hulett MD, Hogarth PM. Molecular basis of Fc Receptor function. Adv Immunol 1994; 57:1-127.
4. Orloff DG, Ra C, Frank SJ et al. The zeta and eta chains of the T cell receptor and the gamma chain of Fc receptors form a family of disulfide-linked dimers. Nature 1990; 347:189-191.
5. Kurosaki T, Gander I, Wirthmueller U et al. The β subunit of the FcεRI is associated with the FcγRIII on mast cells. J Exp Med 1992; 175:447-451.
6. Kinet JP, Blank U, Ra C et al. Isolation and characterization of cDNAs coding for the β subunit of the high-affinity receptor for immunoglobulin E. Proc Natl Acad Sci USA 1988; 85:6483-6487.
7. Reth MG. Antigen receptor tail clue. Nature 1989; 338:383-384.
8. Cambier JC. New nomenclature for the Reth motif (or ARH1/TAM/ARAM/YXXL). Immunol Today 1994; 16:110-110.
9. Heldin CH, Dimerization of cell surface receptors in signal transduction. Cell 1995; 80:213-223.

10. Daëron M, Bonnerot C, Latour S et al. Murine recombinant FcγRIII, but not FcγRII, trigger serotonin release in rat basophilic leukemia cells. J Immunol 1992; 149:1365-1373.

11. Bonnerot C, Amigorena S, Choquet D et al. Role of associated γ chain in tyrosine kinase activation via murine FcγRIII. EMBO J 1992; 11: 2747-2757.

12. Hibbs ML, Walker ID, Kirszbaum L et al. The murine Fc receptor for immunoglobulin: purification, partial amino acid sequence, and isolation of cDNA clones. Proc Natl Acad Sci USA 1986; 83:6980-6984.

13. Lewis VA, Koch T, Plutner H et al. A complementary DNA clone for a macrophage-lymphocyte Fc receptor. Nature 1986; 324:372.

14. Ravetch JV, Luster AD, Weinshank R et al. Structural heterogeneity and functional domains of murine Immunoglobulin G Fc receptors. Science 1986; 234:718-725.

15. Stuart SG, Simister NE, Clarkson SB et al. IgG Fc receptor (hFcRII ; CD32) exists as multiple isoforms in macrophages, lymphocytes and IgG-transporting placental epithelium. EMBO J 1989; 8:3657-3666.

16. Hibbs ML, Bonadonna L, Scott BM et al. Molecular cloning of a human Immunoglobulin G Fc receptor. Proc Natl Acad Sci USA 1988; 85:2240-2244.

17. Brooks DG, Qiu WQ, Luster AD et al. Structure and expression of human IgG FcRII (CD32). Functional heterogeneity is encoded by the alternatively spliced products of multiple genes. J Exp Med 1989; 170:1369-1386.

18. Hogarth PM, Witort E, Hulett MD et al. Structure of the mouse βFcγ receptor II gene. J Immunol 1991; 146:369-376.

19. Latour S, Fridman WH, Daëron M. Identification, molecular cloning, biological properties and tissue distribution of a novel isoform of murine low-affinity IgG receptor homologous to human FcγRIIB1. J Immunol 1996; in press.

20. Miettinen HM, Rose JK, Mellman I. Fc receptor isoforms exhibit distinct abilities for coated pit localization as a result of cytoplasmic domain heterogeneity. Cell 1989; 58:317-327.

21. Daëron M, Malbec O, Latour S et al. Distinct intracytoplasmic sequences are required for endocytosis and phagocytosis via murine FcγRII in mast cells. Intern Immunol 1993; 5:1393-1401.

22. Van den Herik-Oudijk IE, Capel PJA, Van der Bruggen T et al. Identification of signalling motifs within human FcγIIA and FcγRIIb isoforms. Blood 1995; 85:2202-2211.

23. Miettinen HM, Matter K, Hunziker W et al. Fc receptor endocytosis is controlled by a cytoplasmic domain determinant that actively prevents coated pit localization. J Cell Biol 1992; 116:875.

24. Amigorena S, Bonnerot C, Drake J et al. Cytoplasmic domain heterogeneity and functions of IgG Fc receptors in B lymphocytes. Science 1992; 256:1808-1812.

25. Amigorena S, Bonnerot C, Choquet D et al. FcγRII expression in resting and activated B lymphocytes. Eur J Immunol 1989; 19:1379-1385.

26. Néauport-Sautès C, Dupuis D, Fridman WH. Specificity of Fc receptors of activated T cells. Relation with released immunoglobulin-binding factor. EurJ Immunol 1975; 5:849-854.

27. Daëron M, Latour S, Malbec O et al. The same tyrosine-based inhibition motif, in the intracytoplasmic domain of FcγRIIB, regulates negatively BCR-, TCR-, and FcR-dependent cell activation. Immunity 1995; 3:635-646.

28. Benhamou M, Bonnerot C, Fridman WH et al. Molecular heterogeneity of murine mast cell Fcγ receptors. J Immunol 1990; 144:3071-3077.

29. Esposito-Farese M-E, Sautès C, de la Salle H et al. Membrane and soluble FcγRII/III modulate the antigen-presenting capacity of murine dendritic epidermal Langerhans cells for IgG-complexed antigens. J Immunol 1995; 154:1725-1736.

30. Daëron M, Bonnerot C, Latour S et al. The murine αFcγR gene product : identification, expression and regulation. Mol Immunol 1990; 27:1181-1188.

31. Möller G, Wigzell H. Antibody synthesis at the cellular level. Antibody-induced suppression of 19S and 7S antibody response. J Exp Med 1965; 121:969.

32. Henry C, Jerne NK. Competition of 19S and 7S antigen receptors in the regulation of the primary immune response. J Exp Med 1968; 128: 133-145.

33. Uhr JW, Möller G. Regulatory effect of antibody on the immune response. Adv Immunol 1968; 8:81.

34. Safford Jr JW, Tokuda S. Antibody-mediated suppression of the immune response: effect on the development of immunological memory. J Immunol 1971; 107:1213.

35. Shek PN, Dubiski S. Allotypic suppression in rabbits: competition for target cell receptors between isologous and heterologous antibody and between native antibody and antibody fragments. J Immunol 1975; 114:621.

36. Kohler H, Richardson BC, Smyk S. Immune response to phosphorylcholine. IV. Comparison of homologous and isologous anti-idiotypic antibody. J Immunol 1978; 120:233-238.

37. Sinclair NRS, Chan PL. Regulation of the immune response. IV. The role of the Fc-fragment in feedback inhibition by antibody. Adv Exp Med Biol 1971; 12:609-615.

38. Sinclair NRS, Lees RK, Abrahams S et al.Regulation of the immune response. X. Antigen-antibody complex inactivation of cells involved in adoptive transfer. J Immunol 1974; 113:1493.

39. Tew JG, Greene EJ, Makovski MH. In vitro evidence indicating a role for the Fc region of IgG in the mechanism for the long-term maintenace of antibody levels in vivo. Cell. Immunol 1976; 25:141.

40. Kohler H, Richardson B, Rowley DA et al. Immune response to phosphorylcholine. III. Requirement of the Fc portion and equal effectiveness of IgG subclasses in anti-receptor antibody-induced suppression. J Immunol 1977; 119:1979-1986.

41. Stockinger B, Lemmel EM. Fc Receptor dependency of antibody-mediated feedback regulation: on the mechanism of inhibition. Cell Immunol 1978; 40:395-403.

42. Sidman CL, Unanue ER. Requirements for mitogenic stimulation of murine B cells by soluble anti-IgM antibodies. J Immunol 1979; 122:406-413.

43. Sidman CL, Unanue ER. Control ob B lymphocyte function I. Inactivation of mitogenesis by interactions with surface immunoglobulin and Fc-Receptor molecules. J Exp Med 1976; 144:882-896.

44. Unkeless JC. Characterization of monoclonal antibody directed against mouse macrophage and lymphocyte Fc receptors. J Exp Med 1979; 150:580-596.

45. Phillips NE, Parker DC. Cross-linking of B lymphocyte Fcγ receptors and membrane immunoglobulin inhibits anti-immunoglobulin-induced blastogenesis. J Immunol 1984; 132:627-632.

46. Jones B, Tite JP, Janeway Jr. CA. Different phenotypic variants of the mouse B cell tumor A20/2J are selected by antigen- and mitogen-triggered cytotoxicity of L3T4-positive, I-A-restricted T cell clones. J Immunol 1986; 136:348-356.

47. Van den Herik-Oudijk IE, Westerdaal NAC, Henriquez NV et al. Functional analysis of human FcγRII (CD32) isoforms expressed in B lymphocytes. J Immunol 1994; 152:574-585.

48. Katz HR, Arm JP, Benson AC et al. Maturation-related changes in the expression of FcγRII and FcγRIII on mouse mast cells derived in vitro and in vivo. J Immunol 1990; 145:3412-3417.

49. Barsumian EL, Isersky C, Petrino MG et al. IgE-induced histamine release from rat basophilic leukemia cell lines: isolation of releasing and nonreleasing clones. Eur J Immunol 1981; 11:317.

50. Daëron M, Malbec O, Latour S et al. Regulation of high-affinity IgE receptor-mediated mast cell activation by murine low-affinity IgG receptors. J Clin Invest 1995; 95:577-585.

51. Wegener A-M, Letourneur F, Hoeveler A et al. The T cell receptor/CD3 complex is composed of at least two autonomous transduction modules. Cell 1992; 68:83-95.

52. Muta T, Kurosaki T, Misulovin Z et al. A 13-amino-acid motif in the cytoplasmic domain of FcγRIIB modulates B cell receptor signalling. Nature 1994; 368:70-73.

53. Perussia B, Tutt MM, Qui WQ et al. Murine natural killer cells express functional Fcγ receptor II encoded by the FcγRα gene. J Exp Med 1989; 170:73-86.

54. Bonnerot C, Amigorena S, Fridman WH et al. Unmethylation of specific sites in the 5' region is critical for the expression of murine αFcγR gene. J Immunol 1990; 144:323-328.

55. Colonna M, Samaridis J. Cloning of immunoglobulin-superfamily members associated with HLA-C and HLA-B recognition by human natural killer cells. Science 1995; 268:405-408.

56. Wagtmann N, Biassoni R, Cantoni C et al. Molecular clones of the p58 NK cell receptor reveal immunoglobulin-related molecules with diversity in both the extra- and intracellular domains. Immunity 1995; 2:439-449.

57. d'Andrea A, Chang C, Franz-Bacon K et al. Molecular cloning of NKB1, a natural killer cell receptor for HLA-B allotypes. J Immunol 1995; 155:2306-2310.

58. Nakajima H, Tomiyama H, Takiguchi M. Inhibition of γδ T cell recognition by receptors for MHC class I molecules. J Immunol 1995; 155:4139-4142.

59. Burshtyn DN, Scharenberg AM, Wagtmann N et al. Recruitment of tyrosine phosphatase HCP by the killer cell inhibitory receptor. Immunity 1996; 4:77-85.

60. Olcese L, Lang P, Vély F et al. Human and mouse killer-cell inhibitory receptors recruit PTP1C and PTP1D protein tyrosine phosphatases. J Immunol 1996; 156:in press.

61. Choquet D, Partiseti M, Amigorena S et al. Cross-linking of IgG receptors inhibits membrane immunoglobulin-stimulated calcium influx in B lymphocytes. J. Cell Biol 1993; 121:355-363.

62. Diegel ML, Rankin BM, Bolen JB et al. Cross-linking of Fcγ Receptor to surface immunoglobulin on B cells provides an inhibitory signal that closes the plasma membrane calcium channel. J Biol Chem 1994; 15: 11407-11416.

63. D'Ambrosio D, Hippen KH, Minskoff SA et al. Recruitment and activation of PTP1C in negative regulation of antigen receptor signaling by FcgRIIB1. Science 1995; 268:293-296.

64. Claphman DE. Calcium signaling. Cell 1995; 80:259-268.

65. Divecha N, Irvine RF. Phospholipid signaling. Cell 1995; 80:269-278.

66. Cohen GB, Ren R, Baltimore D. Modular binding domains in signal transduction proteins. Cell 1995; 80:237-248.

67. Adachi M, Fischer EH, Ihle J et al. Mammalian SH2-containing protein tyrosine phosphatases. Cell 1996; 85:15-15.

68a. Yi T, Cleveland JL, Ihle JN. Protein tyrosine phosphatase containing SH2 domains: characterization, preferential expression in hematopoietic cells, and localization to human chromosome 12p12-p13. Mol Cell Biol 1992; 12:836-846.

68b. Ono M, Bolland S, Tempst P et al. Role of the inositol phosphatase SHIP in negative regulation of the immune system by the receptor FcγRIIB. Nature 1996; 383:263-266.

69. Van Toorenenbergen AW, Aalberse RC. IgG4 and passive sensitization of basophil leukocytes. Int Arch Allergy Appl Immunol 1981; 65:432-440.

70. Anselmino LM, Perussia B, Thomas LL. Human basophils selectively express the FcγRII (CDw32) subtype of IgG receptor. J. Allergy Clin. Immunol 1989; 84:907-914.

71. Frank SJ, Niklinska BB, Orloff DG et al. Structural mutations of the T cell receptor ζ chain and its role in T cell activation. Science 1990; 249:174-177.

72. Ljunggren HG, Kärre K. Host resistance directed selectively against H-2-deficient lymphoma variants: analysis of the mechanism. J Exp Med 1985; 162:1745-1759.

73. Takai T, Ono M, Hikida M et al. Augmented humoral and anaphylactic responses in FcγRII-deficient mice. Nature 1996; 379:346-349.

74. Greene MC, Shultz LD. Motheaten, an immunodeficient mutant of the mouse. Genetics and pathology. J Hered 1975; 66:250-258.

75. Shultz LD, Schweitzer PA, Rajan TV et al. Mutations at the murine motheaten locus are within the hematopoietic cell protein-tyrosine phosphatase Hcph) gene. Cell 1993; 73:1445-1454.

76. Gleich GJ, Zimmermann EM, Henderson LL et al. Effect of immunotherapy on immunoglobulin E and immunoglobulin G antibodies to ragweed antigens: a six-year prospective study. J Allergy Clin Immunol 1978; 62:261.

77. Machiels JJ, Somville MA, Jacquemin MG et al. Allergen-antibody complexes can efficiently prevent seasonal rhinitis and asthma in grass pollen hypersensitive patients. Allergy 1991; 46:335-348.

78. Machiels JJ, Lebrun PM, Jacquemin MG et al. Significant reduction of nonspecific bronchial reactivity in patients with Dermatophagides pteronyssinus-sensitive allergic asthma under therapy with allergen-antibody complexes. Am Rev Respir Dis 1993; 147:1407-1412.

79. Jutel M, Pichler WJ, Skrbic D et al. Bee vebom immunotherapy results in decrease of IL-4 and IL-5 and increase of IFN-γ secretion in specific allergen-stimulated T cell cultures. J Immunol 1995; 154:4187-4194.

80. Kaliss N. Immunological enhancement of tumor homografts in mice. A review. Cancer Res 1958; 18:992-1035.

81. Voisin GA. Immunological facilitation, a broadening of the concept of the enhancement phenomenon. Progr Allergy 1971; 15:328-375.

82. Capel PJA, Tamboer WPM, De Waal RMW et al. Passive enhancement of skin grafts by alloantibodies is Fc dependent. J Immunol 1979; 122:421-429.

83. Leclerc JC, Plater C, Fridman WH. The role of Fc receptors (FcR) on thymus-derived lymphocytes. I. Presence of FcR on cytotoxic lymphocytes and absence of direct role in cytotoxicity. Eur J Immunol 1977; 7:543-548.

84. Avrameas S, Guilbert B, Dighiero G. Natural antibodies against actin, tubulin, myoglobin, thyroglobulin, fetuin, albumin and transferin are present in normal human sera and monoclonal immunoglobulins from multiple myeloma and Waldeström macroglobulinemia. Ann Immunol (Inst Pasteur) 1981; 132C:231-240.

85. Hall PD. Immunomodulation with intraveinous immunoglobulin. Pharmacotherapy 1993; 13:564-573.

86. Kazatchkine M, Dietrich G, Hurez V et al. V region-mediated selection of autoreactive repertoires by intravenous immunoglobulin (i.v.Ig). Immunol Rev 1994; 139:79-107.
87. Pribluda VS, Pribluda C, Metzger H. Transphophorylation as the mechanism by which the high affinity receptor for IgE is phosphorylated upon aggregation. Proc Natl Acad Sci USA 1994; 91:11246-11250.

CHAPTER 5

INTERNALIZATION THROUGH RECEPTORS FOR IMMUNOGLOBULINS

Sebastian Amigorena

INTRODUCTION

All cells eat part of their environment by internalization. Mammalian cells have developed a number of different modes of internalization, which include pinocytosis (the internalization of liquid medium), endocytosis (receptor mediated internalization of soluble molecules) and phagocytosis (internalization of large particles).[1] Although the primary function of internalization is nutrition, differentiated cell types have developed specific adaptations of the endocytic process which serve a variety of biological functions. Striking examples are synaptic vesicles in neuronal cells or transcytotic vesicles in epithelia, which represent specialized endocytic vesicles playing crucial roles in synaptic transmission or transport across epithelial barriers, respectively. Cells of the immune system have also developed specialization of their endocytic pathway corresponding to specialized functions related to internalization (like antigen presentation or cytotoxic defense against viral, bacterial and protozoan pathogens). Importantly, receptors for the Fc region of immunoglobulins (FcRs) are involved in most of internalization-related aspects of immune responses. Actually, internalization through FcRs plays a major role in antigen presentation, in killing of bacteria and parasites after phagocytosis, in the clearance of immune complexes from the serum, in the downregulation of receptors at the surface of immune competent cells and, finally, in the transport of antibodies across epithelia.

Cell-Mediated Effects of Immunoglobulins, edited by Wolf Herman Fridman and Catherine Sautès. © 1997 R.G. Landes Company.

Understanding how molecules and particles enter cells through specific receptors has been a subject of intense investigation for the last 20 years and FcRs have been one of the important experimental models used in these studies. Internalization through FcRs assumes two main pathways, endocytosis and phagocytosis.[2,3] These two internalization modes are distinguished by at least two main criteria: (1) the size of the internalized particle, and (2) the cytosolic machinery responsible for the internalization process.[4] Endocytosis allows receptor-mediated internalization of soluble ligands through coated pits and coated vesicles. During phagocytosis, large particular ligands are engulfed by plasma membrane and internalized into phogocytic vacuoles. An usual way to distinguish these two internalization pathways is the requirement of an intact actin cytoskeleton for phagocytosis, whereas endocytosis is in most cases independent of actin.[4]

Whatever its internalization path, the fate of the intracellular ligands is also determined by specific receptors. After endocytosis, FcRs determine recycling from endosomes to the cell surface, or their transport to degradative lysosomal compartments.[5,6] After phagocytosis, FcRs trigger the killing of internalized parasites by inducing the fusion of parasitoforous vacuoles with lysosomes.[7] Finally, in polarized cells, FcR determine transport to apical and basolateral surfaces, thus mediating transport of Ig or immune complexes through epithelial barriers.[8,9]

In the last 5 years, the expression of recombinant FcRs in model cell lines helped define the role of particular receptors in various cell functions. Extensive mutagenesis of FcR cytosolic regions led to a detailed mapping of the molecular motifs involved in both ligand internalization and intracellular transport. Modifications of these motifs allowed manipulation of the subcellular localization of FcRs and helped understanding the basis of important cellular functions, as antigen presentation or transcytosis.

Depending on their biological functions and their pattern of expression, particular FcRs were more specifically studied for certain of these functions. For example, receptors for IgG have been extensively used as models to study the endocytosis of immune complexes, antigen presentation and phagocytosis of bacteria and parasites. Receptors for polymeric Ig (pIg-R) have been used to analyze transport across epithelia.

In this chapter, we will sum up our current understanding of FcR endocytosis, phagocytosis and transcytosis capacities. We will review the abilities of particular FcRs to mediate ligand internalization and review the analysis of the molecular motifs involved in various modes of FcR internalization. We will also stress how the expression of particular FcRs relates to the biological functions of the cells and define the contribution of studies on FcRs to the analysis of fundamental mechanisms of intracellular transport.

ENDOCYTOSIS

WHERE TO GO: THE ENDOCYTIC PATHWAY AND FcRs

Endocytosis concerns soluble ligands which may or may not bind to specific receptors at the cell surface. Endocytosis of soluble ligands which do not bind to specific membrane receptors is generally referred to as « pinocytosis » or « macropinocytosis » depending on the size of the endocytic vesicles where the internalized material is detected.[1]

In the case of ligands which do bind to membrane receptors, the receptors concentrate in specialized regions of the plasma membrane, the coated pits (Fig. 5.1). Some receptors are constitutely clustered in coated pits and efficiently internalized (like transferrin receptor, TfnR or low density lipoprotein receptor, LDLR), whereas other receptors require binding to their specific ligands to localize to coated pits (that is the case of the epidermal growth factor receptor, EGFR, for example). After clustering in coated pits, receptors are internalized by the budding of coated vesicles from the plasma membrane (Fig. 5.1). These vesicles become competent to fuse with early endosomes after dissociation of the clathrin coat.[1]

Once in early endosomes, internalized receptor-ligand complexes are exposed to a slightly acidic milieu (pH 6-6.5) (Fig. 5.1). At this pH, certain ligands dissociate from their receptors while others do not. LDL, for example, dissociates from its receptor, whereas Tfn or immune complexes do not. This sensitivity of ligand-receptor interaction will determine the subsequent fate of the endocytosed molecules. Receptors, in most cases (Tfn-R or LDLR, for example), recycle to the cell surface after sorting in early endosomes. Some ligands, like LDL, are delivered to late endosomes and lysosomes to be degraded, after dissociation from their receptors (Fig. 5.1). Tfn loses its bound iron molecules at endosomal pH before recycling to the plasma membrane.[1]

FcγR-mediated internalization was first characterized on macrophage cell lines.[2] In the absence of ligand, FcγR were internalized into early endosomes, and recycled to the cell surface.[6] Multivalent ligands (like IgG-containing immune complexes) were also rapidly internalized, but their rate of recycling was extremely low. Instead, multivalent ligands were efficiently delivered to lysosomes and degraded (Fig. 5.1).[5] Therefore, aggregation may influence receptor-ligand sorting in endosomes and determine transport to degradative lysosomal compartments or recycling to the plasma membrane.

These efforts towards understanding FcR's endocytic capabilities preceded the molecular characterization of the actual receptors involved in internalization. In 1987, the cloning of cDNAs encoding various receptors for IgG showed that members of the FcR family had relatively well conserved extracellular domains and very diverse cytosolic regions[10,11] (see chapter 2). This is even more striking in the case of low affinity receptors for IgG (type II and III) (Fig. 5.2). In mice, two isoforms of

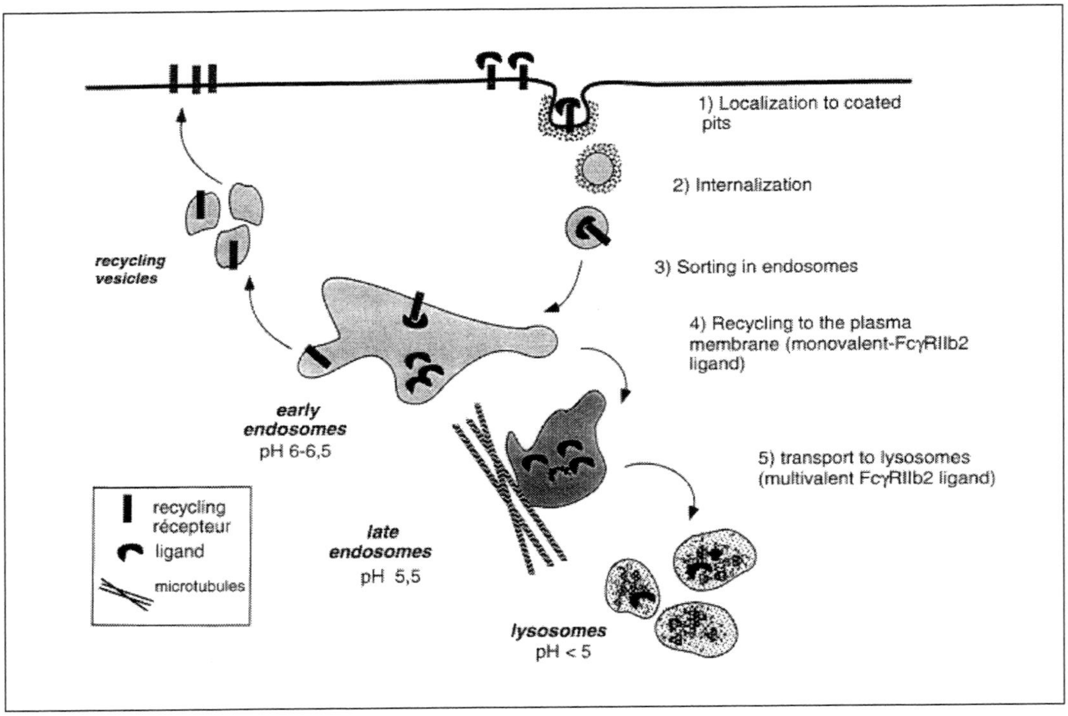

Fig. 5.1. Functional organization of the endocytic pathway. Upon localization to clathrin coated pits, receptor-ligand complexes are internalized into coated vesicles, which fuse with early endosomes after uncoating. Exposure to acidic pH found in early endosomes induces the dissociation of most receptor-ligand complexes, but not of FcR-Ig interactions. Early endosomes are sorting organelles: membrane proteins can either be recycled to the cell surface or be delivered, via a microtubule-dependent transport step, to late endosomes and lysosomes. In the case of many receptor-ligand complexes, like LDL-LDLR, receptors recycle while free ligands reach lysosomes and are degraded. For FcγR, multivalent ligands follow the degradation route, whereas monovalent ligands or ligand-free receptors recycle.

type II FcγR, mFcγRIIb1 and mFcγRIIb2, are generated by alternative splicing of sequences coding for 47 amino acid residues of the cytoplasmic domain (Fig. 5.2).[12] In humans, there also exists a third form of FcγRII, hFcγRIIA, composed of extracellular and transmembrane domains homologous to the mFcγRII, and a completely different cytoplasmic tail.[12]

Type III FcγR (FcγRIII) is a multimeric receptor composed of three subunits: an α chain and a homodimer of γ chains, which was initially described as part of FcεRI[12] (Fig. 5.2). FcγRIII α chain is the ligand binding subunit; it is 95% homologous to FcγRII in the extracellular domains and less than 20% homologous to FcγRII in the transmembrane and the cytoplasmic regions. In humans, a glycosyl-phosphatidylinositol (GPI)-linked form of hFcγRIII (hFcγRIIIB), which is not associated with γ chains also exists.[12] Besides FcεRI and FcγRIII, high affinity FcγRI and FcαRs also associate with γ chains (see chapter 2).[9]

The γ chain shares sequence homology with the homodimeric ζ chain of the antigen receptor of T lymphocytes (TcR). Importantly, γ chains also contain immunoreceptor tyrosine-containing activation motifs (ITAMs),[13] which are involved in many of FcR's biological functions, including internalization.[14,15] Besides those FcR which associate to γ chains (FcεRI, FcγRI, FcγRIII and FcαR), hFcγRIIA also contains an ITAM in the cytosolic domain of its ligand-binding subunit.[16] The other FcγRII members of the family, contain a structurally-related motif, immuno-receptor tyrosine-containing inhibition motifs (ITIM), involved in the downregulation of cell activation through other immunoreceptors.[17-19]

Thus, FcRs represent a complex family of membrane proteins with a peculiar structural characteristic: high molecular variability within the cytosolic domains and relatively low variability in the extracellular regions, within each of the subfamilies specific for particular Ig isotypes. This high variability in the cytosolic domains results from both the existence of several genes for a particular type of receptor, from alternative splicing of exons encoding part of the cytosolic domains and/or from the association of ligand-binding chains to various non-ligand binding subunits.

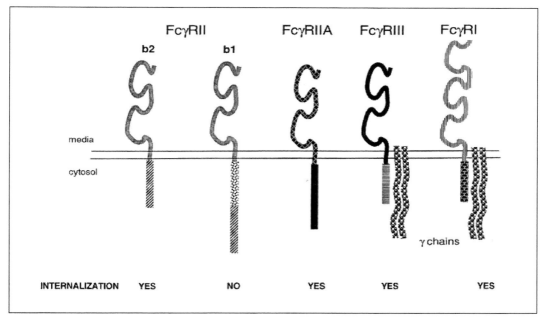

Fig. 5.2. Endocytosis of murine IgG receptors. Types I and III receptors for IgG associate to γ chains, whereas the two type II receptors, b1 and b2 do not. Like other receptors which associate to γ chains (which also include FcεRI and FcαRs), FcγRI and FcγRIII are efficiently internalized due to the strong internalization signal within the ITAM present in the γ chain (see Table 5.1). The two tyrosine residues present in this ITAM are both required for an efficient internalization of the receptor. FcγRIIb1 and b2 also contain an internalization signal based on leucine residues. The 47 amino acid insertion in the cytoplasmic domain of FcγRIIb1 inactivates this internalization signal, thus preventing the endocytosis of immune complexes by this receptor.

To understand the functional relevance of this diversity, a number of groups undertook the expression of single recombinant FcRs or of chimeric receptors containing the cytosolic domains of associated chains, in model cell lines. Over the past few years, this approach led to the attribution of particular biological functions to certain receptors. By mutagenesis of the cytosolic domains, this approach also permitted assigning particular amino acid domains within the receptor's cytosolic tails to defined functions. The definition of the cytosolic effectors interacting with these domains represents the next step towards understanding the molecular basis of FcR functions.

WHO GOES: ENDOCYTOSIS BY FCRS

The first characterization of the internalization capabilities of a recombinant FcR was performed by the group of I. Mellman.[20] The two isoforms of mFcγRII, b1 and b2, were expressed in fibroblasts. Only mFcγRIIb2, and not mFcγRIIb1, efficiently localized to coated pits and mediated the endocytosis of immune-complexes. mFcγRIIb1 was not actively endocytosed, due to the cytoplasmic insertion present in its cytoplasmic domain (Fig. 5.2). mFcγRIIb1 was retained at the cell surface by being excluded from coated regions of the plasma membrane. This effect was due to the 47 amino acid insertion, since the addition of this sequence to the COOH-terminus of the mFcγRIIb2 isoform blocked endocytosis.

In addition, ligand-induced crosslinking of mFcγRIIb1 by immune complexes triggered the interaction of the receptor with cytoskeleton.[21] Consistently, when expressed in B lymphoma cells, mFcγRIIb1 crosslinking lead to the formation of typical cap structures at one pole of the cell.[17] These observations suggested that crosslinking induced the interaction of mFcγRIIb1 with the cytoskeleton through the 47 amino acids insertion. This association with the cytoskeleton could account for the inability of mFcγRIIb1 to endocytose immunocomplexes. This possibility, however, was discarded by the finding that within mFcγRIIb1's 47 amino acid insertion, the amino-terminal half which was sufficient to block internalization did to promote association to cytoskeleton (K. Matter and I. Mellman, personal communication). Interestingly, the human homologue of the mFcγRIIb1 only contains this region of the insert.

The inhibition of endocytosis by the mFcγRIIb1-insertion was only effective for mFcγRIIb2 mediated internalization: no effect was observed on endocytosis promoted by the LDLR (K. Matter, personal communication) or by FcεRI-associated γ chain (C. Bonnerot, personal communication).

Thus, the only positive function associated with the mFcγRIIb1 insert itself was an increased ability of mFcγRII to cap in response to cross-linking.[17] Since the insert may increase the propensity of the receptors to associate with a detergent-insoluble fraction in fibroblasts, this function can be understood as capping in lymphocytes is thought to rely on direct or indirect attachment of membrane proteins to the underlying cytoskeleton. Increased capping efficiency may be important for FcγRII function on B cells, particularly during B cell activation. Capping of sIg leads not only to the co-capping of mFcγRII, but also includes several other molecules that may be involved in the activation process; these include MHC class II molecules, CD19, C3d receptor, and cytoplasmic H-*ras*.[14,17] Thus, the increased propensity of mFcγRIIb1 to cap may ensure that the receptor is distributed together with the molecules whose activities it must regulate following sIg-FcγRII crosslinking. In addition, it is conceivable that the possible interaction of the mFcγRIIb1 cytoplasmic tail with the cytoskeleton may help to organize the activation complex itself, perhaps by providing an indirect link between molecules whose cytoplasmic domains are incapable of such interactions.

Internalization of mFcγRIII was also analyzed in this system.[22] mFcγRIII α and γ chains co-expressed in B lymphoma cells were fully competent to endocytose immune complexes (Fig. 5.2). Surprisingly, deletion of the cytosolic domain of mFcγRIII-ligand binding α subunit did not alter internalization. In contrast, when the α chain was expressed in the absence of γ chain, no internalization of immune complexes was observed. Other FcR, including hFcγRIIA[23] and hFcεRI,[24] also mediate efficient endocytosis.

READY, ...GO!: SIGNALS FOR ENDOCYTOSIS IN FCRS

The remarkable efficiency of all of these receptors for endocytosis of immune complexes raised the question of the signals for their internalization. The first step of endocytosis, concentration of receptors in coated pits, is due to the binding of their cytosolic tails to a complex of proteins called "adaptors".[25] These adaptors allow the formation of a "coat" on the cytosolic side of the plasma membrane, through the binding of yet another complex of proteins, which in the case of receptor endocytosis, is clathrin. The coated pits form coated vesicles, due to a fission event, resulting in the actual internalization of the receptors, their ligands and a certain volume of liquid medium[26] (Fig. 5.1).

Accumulation of receptors into coated pits is due to specific signals present in the cytosolic domains of the receptors. These signals have been extensively characterized in the case of a number of receptors, including TfnR, LDLR, as well as number of FcRs.[27] Internalization signals generally consist of 4-6 amino acids. Most of these signals depend on an aromatic residue in the context of a tight turn. However, double leucine internalization motifs were also described in a number of membrane receptors.[28]

In the case of FcRs, FcγRII have been most extensively studied for their endocytic capabilities. The internalization signal of mFcγRIIb2 was mapped to the proximal cytoplasmic sequence EAENTITYSLLKH (amino acids 18 to 31 of the cytosolic domain, Table 5.1). Deletion

Table 5.1. Endocytosis signals in FcγR

FcR	SIGNAL
FcγRIIb2	EAENTITYS[LL]KHDEETEHDYQNHI
FcγRIII* FcγRI*	KADAV[Y]TGLNTRSQET[Y]ETLKHE
FcγRIIA	TADGGYMTLNPRAPTDDDNKIYLTL

*the internalization signal in these FcR is present in the associated γ chain

of the distal portion of the cytoplasmic tail,[20,17] containing the other tyrosine residue (in position 39), or mutation of this tyrosine, did not modify the ability of the receptor to be internalized.[29] Replacement of the first tyrosine residue (26) by a glycine or mutation of the two leucines in this sequence of thirteen amino acids completely abolished ligand-induced mFcγRIIb2 internalization in RBL cells[29] and in B lymphocytes (unpublished results). However, replacement of the same tyrosine by alanine (in mFcγRIIb2)[30,31] or by phenylalanine (in hFcγRIIb2)[32] had no effect on endocytosis. In every case, mutations affecting the double leucine completely blocked endocytosis.

Another endocytosis-competent FcR is hFcγRIIA. This receptor contains a di-tyrosine based activation motif in its own cytoplasmic domain and effectively induces activation of tyrosine kinases and Ca^{2+} mobilization.[33] FcγRIIA efficiently internalized ligands when expressed in either macrophages or CHO cells.[23]

By constructing chimeric receptors, in which the extracellular and transmembrane domains of FcγRIII were fused to the cytoplasmic tail of the γ chain, we showed that the internalization signal of mFcγRIII is located in the cytoplasmic domain of the associated-subunit (Table 5.2), which also contains two tyrosine residues involved in triggering cell activation.[22] Exchange of either tyrosine residue inhibited the endocytosis of the chimeras. Thus, mFcγRIII has a di-tyrosine-based internalization signal located in the cytoplasmic domain of the associated γ subunit (Table 5.2). The cytoplasmic domains of the structurally related CD3 ε chains also contain a tyrosine-based motif and a di-leucine motif which independently mediated the endocytosis of chimeric

Table 5.2. Phagocytosis signals in FcγR

FcR	SIGNAL
FcγRIIb2	**EAENTITYSLLKHDEETEHDYQNHI**
FcγRIII* FcγRI*	**KADAVYTGLNTRSQETYETLKHE**
FcγRIIA	**TADGGYMTLNPRAPTDDDNKIYLTL**

*the phagocytosis signal in these FcR is present in the associated γ chain

proteins constructed with the extracellular and transmembrane domains of the Tac antigen.[28] The presence of this di-tyrosine-based motif in the cytoplasmic domain of other associated chains, like the ζ chain of the TcR or the α and β chains associated with sIg raises the possibility of a general role of these associated subunits in receptor internalization.

The differences in the internalization signals in mFcγRIIb2 and mFcγRIII suggest that different cytoplasmic effectors might be involved in the uptake of immune complexes via these two receptors. The internalization signal in the cytoplasmic tail of mFcγRIIb2 involves only one tyrosine residue although two tyrosine residues are present in this portion of the receptor. This situation is similar to the LDL receptor which contains three tyrosine residues in its cytoplasmic domain, but only one of them is involved in receptor internalization.[8]

Endocytosis via mFcγRIII is the first example of an internalization signal which is not located in the cytoplasmic tail of the ligand-binding chain, but in the cytoplasmic domain of an associated chain. Another particularity of this receptor is the requirement of two tyrosine residues for ligand internalization. This exact same sequence is also involved in the triggering of cell activation by mediating interactions with tyrosine protein kinases.[12] Studies of EGF receptor internalization suggests that endocytosis and cell activation might be related. The cytoplasmic domain of the EGF receptor also contains an internalization signal and protein kinase activity. Mutations affecting the tyrosine kinase activity of EGF receptor prevent its lysosomal targeting.[34] The role of PTK in FcγRIII intacellular traffic has not yet been defined.

Thus, two functionally distinct signals have been described for FcR endocytosis: double leucine-containing signals for FcγRIIb2 and ITAMs for all the other endocytic FcRs. Interestingly, no FcR uses "conventional" tight β turn, tyrosine-containing endocytosis signals.

PRESENTING ANTIGENS TO T LYMPHOCYTES

Antigen presentation in the context of major histocompatibility complex (MHC) class II molecules is a crucial step in the initiation of most immune responses. Antigen receptors on helper T lymphocytes only recognize antigen-derived peptides (usually 10-25 amino acids long) which are generated and bind MHC class II within the endocytic pathway of antigen presenting cells.[35,36] These peptides are generally derived from exogenous molecules internalized by the presenting cells. The internalization of antigens actually represents an important step, since at physiological concentrations only antigens internalized by receptor-mediated endocytosis are efficiently presented to T lymphocytes.[37]

In B lymphocytes, antigen internalization by membrane Ig promotes efficient antigen presentation.[38] In most antigen presenting cells, the expression of FcR correlates with the expression of MHC class II molecules (for example in B lymphocytes, macrophages or dendritic cells). Therefore, FcRs represent potential antigen receptors on these "professional" antigen presenting cells.

Ag presentation after internalization through FcRs has been analyzed in a number of different systems. In macrophages and dendritic cells (including Langerhaus cells), FcγR increased the efficiency of antigen presentation to T lymphocytes by promoting the internalization of immune complexes.[39,40] In human myeloid cells the three classes of FcR (I, II and III) enhance the presentation of IgG-complexed antigens.[41] The role of FcγRs in macrophages and dendritic cells is thus analogous to the role of membrane Ig in antigen presentation by B lymphocytes.

An important consequence of the B cell's inability to internalize their FcRs, mFcγRIIb1, was the almost complete inability of complexing antigen with IgG to enhance antigen presentation.[17] Replacing mFcγRIIb1 with an internalization-competent form of the receptor completely reversed this "defect". mFcγRIIb2-expressing B cells efficiently presented antigen-antibody complexes to T cells and directly illustrated the obligatory role of endocytosis in the processing and presentation of exogenous antigen.[17] These results are consistent with previous observations showing that antigen bound to endogenous FcγR on B cells was not efficiently presented, while FcγR on macrophages enhanced presentation.

The fact that B lymphocytes express a FcγRII receptor that is incapable of endocytosis suggests that it is important for B cells to be unable to present FcγR-bound antigens. Normally, antigen presentation by B cells is thought to involve the internalization and processing of antigens following binding to sIg. Nevertheless, mFcγRIIb2 in B cells can enhance the presentation of soluble antigen 1000-fold.[17] By preventing the FcγR-mediated internalization of antigen-antibody complex, presentation by B cells remains antigen-specific thus ensuring the expansion of specific clones of B cells. Should efficient "non-specific" presentation via FcγR be allowed to occur, B cells of any specificity could in principle be stimulated by T cells.

mFcγRIII, when expressed in B lymphoma cells, also mediated efficient antigen presentation to T lymphocytes.[22] As for internalization and cell activation, antigen presentation was dependent on associated γ chains. It is most likely that all FcR which associate with γ chains (FcγRI, FcεRI and FcαR) will turn out to be efficient for antigen presentation.

Thus, all FcRs except FcγRIIb1 are likely to be competent for antigen presentation. Consequently, all cell types expressing both MHC class II and FcRs, except B lymphocytes (where only FcγRIIb1 are expressed), are likely to present antigen-IgG complexes very efficiently. Furthermore, our preliminary results suggest that antigen internalization via two different FcγRs (mFcγRIIb2 and mFcγRIII) induces the antigen presentation to different T hybridoma cells (Bonnerot and Amigorena, unpublished results). Therefore, the particular receptors expressed by antigen presenting cells may determine part of the specificity of the immune response.

Whether or not the primary role of the cytoplasmic domain insert in mFcγRIIb1 is to prevent endocytosis and thereby limit the potential for nonspecific antigen presentation, it is clear that the insertion does not have a specific role in the negative regulation of B cell activation which occurs after co-crosslinking the receptor with sIg.[17] The receptor-dependent abolition of either early (Ca^{2+} influx) or late (cytokine release) responses in B lymphoma cells did not require the insertion itself.[17,42] Instead, a region of the cytoplasmic tail located between residues 18-31 of mFcγRIIb2 appeared to be most important for eliciting these effects. Although this region was also found to be important for coated pit localization, the modulation of sIg induced activation by FcγRII did not correlate with the internalization phenotype of the receptor expressed (both the b1 and b2 isoforms were equally efficient at preventing the activation response). While the mechanism by which the receptor prevents sIg induced activation is still unclear, it is clear that this function is not specific to the FcγRIIb1 isoform. The inability of the FcγRIIb1 isoform to be internalized may indirectly contribute to its regulatory function by maintaining high concentrations of the receptor on the plasma membrane. In macrophages, endocytosis of ligand via mFcγRII leads to lysosomal degradation and downregulation of internalized receptors.[5]

PHAGOCYTOSIS

Phagocytosis of viral, bacterial and protozoan pathogens plays a central role in the early stages of immune responses.[4] Two important

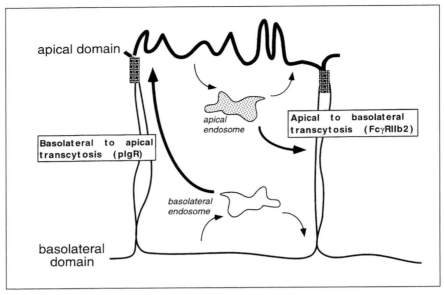

Fig. 5.3. Basic steps of phagocytosis. The first step of phagocytosis, engulfment, requires the presence of IgG over the total circumference of the particles. After engulfment, the particles are internalized by an actin-, protein tyrosine kinase-dependent mechanism, into phagocytic vacuoles. These vacuoles may or may not fuse with lysosomes and become phagolysosomes. In the case of parasitic or bacterial invasion, only after fusion with lysosomes are the pathogens efficiently killed.

phases may be distinguished in the phagocytic process, (i) the engulfment and ingestion of the particles and (ii) its intracellular digestion (Fig. 5.3). FcRs were shown to play important roles in both of these phases.[4]

Engulfment and ingestion require the presence of opsonins (immunoglobulins or complement) at the surface of particles.[4] FcR or complement receptors have to be expressed on the phagocytic cells for a phogocytic process to be efficient. Elegant work from the group of S. Silverstein[43] showed that circumpherential attachment of particles was required for the engulfment, since particles coated with Ig on one half of their surface could not be ingested. After ingestion, the particle resides in a vacuole (the parasitophorous vacuole), which may or may not fuse with lysosomes. This fusion event is a determinant for the subsequent fate of ingested particles (Fig. 5.3). In most cases, only after fusion with lysosomes are the bacteria or parasites killed.[47,44]

The abilities of mFcγRII for phagocytosis of *toxoplasma gondii (T. gondii)*, a protozoan parasite, was analyzed in fibroblastic CHO cells expressing recombinant FcR.[7] *T. gondii* infects mammalian phagocytic and non phagocytic cells. After phagocytosis, the parasite resides and multiplies in vacuoles incompetent to fuse with lysosomes.

However, in macrophages, opsonized *T. gondii* are killed due to the fusion of parasitophorous vacuoles with lysosomes. In transfected CHO cells, both mFcγRIIb1 and mFcγRIIb2, but not a tail-less mFcγRII, mediated the phagocytosis of IgG-coated parasites (Fig. 5.4). Furthermore, both mFcγRII isoforms also induced the efficient fusion of the vacuoles with lysosomes and the killing of the parasites.[7]

A number of other studies also analyzed the abilities of particular receptors for phagocytosis. Odin et al[23] showed that hFcγRIIA were capable of phagocytosing IgG-coated red blood cells when expressed in P388D1 cells but not in CHO cells (Fig. 5.3). They also showed that deletion of the 17 COOH-terminal amino acids of hFcγRIIA cytoplasmic tail differentially affected internalization of different ligands and Ca^{2+} signalling.[23]

Human FcγRI, hFcγRII and hFcγRIIIA were all capable of promoting phagocytosis (Fig. 5.4).[45] Only the GPI-linked hFcγRIIIB on neutrophils was incapable of doing so. Recently, it was shown by Takai et al[46] that macrophages from mice with a disrupted γ chain gene are incapable of phagocytosis of IgG-coated red blood cells. Since the γ chain is associated with FcγRI and III, and not with FcγRII, it was concluded that FcγRI and III must be the phagocytic FcR in macrophages.

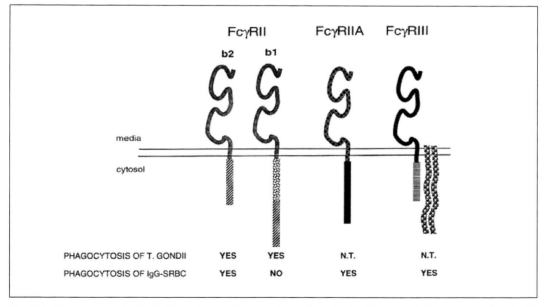

Fig. 5.4 . Phagocytosis by FcgR. The phagocytic capacities of murine FcγR have only been determined with FcγRIIb1 and b2 for the phagocytosis of the protozoan parasite T. gondii in transfected fibroblasts and with mFcγRIIb1, mFcγRIIb2 in transfected RBL cells for the phagocytosis of IgG-coated sheep red blood cells. The two isoforms of mFcγRII mediated the phagocytosis of T. gondii, whereas only the latter was efficient at phagocytosing red blood cells. hFcγRI, hFcγRIIA and hFcγRIII were all capable of phagocytosis of red blood cells when expressed in fibroblastoid cell lines. The signals for phagocytosis by FcγRs are shown in Table 5.2.

The cellular machineries required for endocytosis and phagocytosis are different. Although the cytosolic effectors involved in endocytosis have been described with some detail, very little is known about mechanisms of phagocytosis. It is known, however, that an intact actin cytoskeleton is required for phagocytosis and, in most cases, not for endocytosis.[4,47,48] The internalization event in phagocytosis is therefore mechanistically distinct from the internalization event in the endocytic process.

Several studies have shown that F-actin is recruited to the region of the cell membrane in contact with the particle during FcR-mediated engulfment.[48,49] Furthermore, cytochalasin, which blocks the assembly of F-actin, also inhibits engulfment.[47] Talin, a cytosolic protein known to mediate the interaction of integrins with the cytoskeleton, is also recruited to the sites of engulfment during FcR mediated phagocytosis.[48] In addition, paxillin, a cytoskeleton-associated PTK substrate found in focal adhesions in fibroblasts, was phosphorylated during phagocytosis and co-localized with F-actin beneath nascent phagosomes.[50] These studies suggest a prominent role for the actin cytoskeleton during FcR-mediated phagocytosis, but the actual cytosolic machinery involved in the phagocytic event remains unknown.

Some insight into the activation paths involved in phagocytosis came from studies of FcR-mediated phagocytosis of opsonized erythrocytes. Murine FcγRIIb2, but not b1, expressed in rat basophilic leukemia (RBL) cells, promoted the phagocytosis of anti-FcR-coated red blood cells.[29] Distinct domains of the cytosolic tail were required for endocytosis and phagocytosis.[29] Importantly, the internalization signal in mFcγRIIb2 was not required for phagocytosis, whereas the most C-terminal tyrosine was (this tyrosine was not part of the internalization signal) (Table 5.2).

In contrast to mFcγRII-mediated phagocytosis, which is independent of any associated chain, mFcγRIII-mediated phagocytosis in RBL

Table 5.3. Transcytosis signals in FcγR

FcR	SIGNAL
FcγRIIb2	**EAENTITYSLLKHDEETEHDYQNHI**
pIgR	**NVDRVSIGSYRTDI....KKAKRSSKEEADLAYSAFL**

cells[51] and COS-7[45] cells was determined by the associated γ chain. A similar situation was described for the high affinity hFcγRI, which also requires the associated γ chain for efficient phagocytosis.[52] In these two cases, phagocytosis is dependent on the ITAM present in the cytosolic domain of the γ chain (Table 5.2).

Slightly different is the case of hFcγRIIA, which mediates phagocytosis in the absence of any associated chain, through the ITAM present in its own cytosolic domain (Table 5.2).[23,53] In COS-7 cells, hFcγRII, which contain an incomplete ITAM (with only one YXXL motif), were inefficient for phagocytosis, unless a second YXXL was added.[54]

Interestingly, phagocytosis and endocytosis signals could be physically distinguished in most FcR-mediated phagocytic events. For hFcγRI, only phagocytosis was absolutely dependent on the associated γ chains; the ligand-binding subunit alone was competent for endocytosis.[52] In the case of mFcγRII, the signals for endocytosis and phagocytosis are on different regions of the cytoplasmic domain.[29]

As with most of the other biological functions associated with ITAMs, phagocytosis also depends on tyrosine phosphorylation events, including tyrosine phosphorylation of the ITAM itself.[45] PTK inhibitors reduced phagocytosis and over expression of the PTK Syk, which may be associated with γ chains, enhances phagocytosis.[53,55] Furthermore, Syk becomes phosphorylated during phagocytosis in macrophages.[50] Interestingly, phagocytosis through ITAM-containing receptors depends not only on the tyrosines of the motif (YXXL—YXXL), but also on the two amino acids between the tyrosine and the leucine residues.[55] In contrast, the flanking amino acids, including the residues between the two YXXL did not affect phagocytosis.

However, studies by the group of S. Greenberg showed that intracellular domains of FcR were not absolutely required for coupling to cytoskeleton and phagocytosis.[56] Actually, chimeric molecules containing syk instead of a cytoplasmic domain were fully competent to trigger phagocytosis and actin redistribution in COS cells. A point mutation in syk's catalytic domain blocked phagocytosis.[56] Chimeras containing src family kinases were also inefficient. Therefore, syk kinase plays a central role in triggering phagocytosis.

All together, the studies on phagocytosis by FcγRs suggest the existence of two functionally distinct phagocytosis paths. The first one is mediated by ITAM-negative FcγRs (FcγRII, except FcγRIIA). This phagocytosis process is effective for parasites, which express molecules interacting with the plasma membrane of the phagocytic cells, and may be of capable of transducing activation signals independent of the FcRs. The second phagocytic pathway is only mediated by ITAM-containing FcγRs. This phagocytosis process is also efficient for inert particles, which may not interact with the phagocytic cells by themselves.

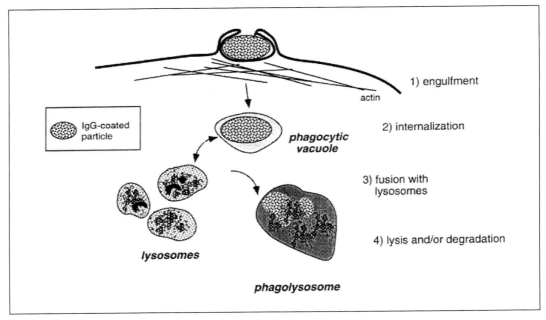

Fig. 5.5. FcR-mediated transcytosis. Epithelial cells have a polarized plasma membrane organization; two distinct domains may be distinguished, the apical and the basolateral membranes. These domains are generated due to the polarized delivery of membrane proteins and lipids to these domains. Proteins are delivered to the apical and basolateral domains either from the secretory pathway or from the endocytic pathway, by transcytosis. Transcytosis requires endocytosis into endosomes, sorting and transport to the opposite membrane. Transcytosis allows transport from the basolateral to the apical surfaces or in the opposite direction

TRANSCYTOSIS

Transport across epithelia is also one of FcR's important roles. Two FcR-dependent transcytosis events have been studied in some detail: IgA transcytosis by the polymeric Ig receptor (pIg-R) and IgG transcytosis by mFcγRII. Secretory IgA is an important component of mucosal protection against pathogens and pIg-R normally mediates IgA and IgM transport across epithelia (from the basolateral to the apical domains) into secretions (Fig. 5.5).[9,57] mFcγRII transport IgG in the opposite direction, from the apical to the basolateral domains, as expected for IgG receptors expressed in placenta (Fig. 5.5).[58]

These two transcytotic events have been analyzed in Madin-Darby canine kidney (MDCK) cells transfected with either pIg-R or mFcγRII. Newly synthesized pIg-R are first transported to the basolateral surface where they bind dimeric IgA.[59] After endocytosis, the pIg-R/dIgA complexes are transported to the apical surface where the receptors are cleaved, thus releasing secretory component/IgA complexes. The signals required for the transport of IgA in polarized cells are contained within the 100 amino acid-long cytosolic tail of the pIg-R.[57] Two regions of the cytosolic tail seem important (Table 5.3). The first determines transport of newly synthesized pIg-R to the basolateral surafce.[59] The second is required for transcytosis.

The signal determining the transport of pIg-R to the basolateral domain is contained within the proximal 17 amino acids and includes a tyrosine-containing tight β-turn motif (Table 5.3).[59] This motif, as is the case of many other basolateral targeting motifs, is also an internalization signal which directs localization into coated pits. However, mutation of the tyrosine in pIg-R internalization motifs did not affect basolateral sorting.[60]

The second transport signal in the pIg-R cytosolic tail includes two serines (664 and 726) which are phosphorylated during transcytosis (Table 5.3).[61] Using a co-culture system where dIgA produced by a hybridoma are transcytosed by recombinant pIg-R expressed in MDCK cells, Hirt et al[62] showed that 1) mutation of serine 664 inhibits transcytosis of unoccupied receptors but after binding of dIgA, the receptors were phosphorylated and capable of transcytosis and 2) mutation of serine 726 drastically reduced transcytosis of dIgA.

The polarized expression of mFcγRIIb1 and mFcγRIIb2 was also analyzed after transfection in the polarized cell line MDCK.[63] The b1 isoform was mainly expressed on the apical surfaces of the cell, whereas mFcγRIIb2 was found on the basolateral membrane. Furthermore, mFcγRIIb2 mediated efficient internalization of immune-complexes and transcytosis between the apical and basolateral membranes (Fig. 5.6). No transcytosis in the opposite direction (basolateral to apical) was observed. Interestingly, transcytosis did not require crosslinking of the receptors, whereas transport to lysosomes, which also occurred in MDCK cells, was only efficient with multivalent ligands.

The signals for basolateral transport of newly synthesized receptors and for transcytosis of immunecomplexes in the mFcγRIIb2 cytosolic domain were mapped by site directed mutagenesis. As for endocytosis, neither of the two tyrosines was required for transcytosis of immune complexes, whereas the double leucine motif was critical (Table 5.3).[63,31]

Studies on transcytosis of Ig by FcRs provides epithelial cells with vectorial transport carriers between the two membrane domains of polarized epithelia: the pIg receptors mediates the secretion of mucosal antibodies by transporting multimeric Ig from the basolateral to the apical domains and mFcγRIIb2 mediates transport in the opposite direction, as would be required for transcellular transport of maternal IgG.

CONCLUSION

Studies on FcR internalization paths have proceeded for more than one century, starting with the pioneer works of Metchnikoff on opsonization and phagocytosis.[4] Since then, studies on FcR-mediated internalization contributed to understanding both the most functional

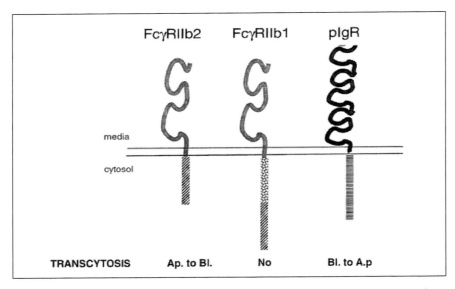

Fig. 5.6. Transcytosis by FcR. Transcytosis by the pIgR allows transport of polymeric Ig from the basolateral (Bl) to the apical (Ap) surfaces, whereas transcytosis by FcγRIIb2 mediates transport of immune complexes in the opposite direction. FcγRIIb1 is incapable of transcytosis, probably due to its inability to internalize ligands. The signals for transcytosis in FcγRIIb2 include the same di-leucine motif required for endocytosis (Table 5.3). The signal for transcytosis of the pIgR includes two serine residues which maybe phosphorylated (see in the text and Table 5.3).

aspects of immunological responses and the most fundamental aspects of cell biology: the works of Z. Cohn on lysosomes and their role in bacterial cytotoxicity, the contributions of S. Silverstein to define the mechanisms of engulfment during phagocytosis, more recently, the work of many groups concerning FcR signal transduction during phagocytosis, the regulation of endocytosis, transcytosis and, finally, various aspects of antigen presentation. These works vividly illustrate how profitable the mixing of two independent scientific fields, immunology and cell biology has been. It also stresses how crucial the consequences of intracellular transport events on the overall behavior of an extremely complex biological system, like the immune system, might be.

REFERENCES

1. Gruenberg J, Maxfield FR. Membrane transport in the endocytic pathway. Current Opinion in Biology 1995; 7:552-563.
2. Mellman I. Endocytosis, membrane recycling and Fc receptor function. In: Pittman, ed. Memrain Recycling, Ciba Foundation Symposium. London: Evard 1982; 35-38.
3. Mellman I. Endocytosis and the entry of intracellular parasites. Infectious Agents and Disease 1993; 2:186-192.

4. Silverstein SC, Greenberg F, Di Viglio F et al. Phagocytosis. In: Paul WE, ed. Fundamental Immunology. New York: Raven Press, 1989; 703.

5. Mellman I, Plutner H. Internalization and degradation of macrophage Fc receptors bound to polyvalent immune complexes. J Cell Biol 1984; 98:1170-1177.

6. Mellman I, Plutner H, Ukkonen P. Internalization and rapid recycling of macrophage Fc receptors tagged with monovalent antireceptor antibody: possible role of a prelysosomal compartment. J Cell Biol 1984; 98:1163-1169.

7. Joiner KA, Fuhrman SA, Miettinen HM et al. Toxoplasma gondii: fusion competence of parasitophorous vacuoles in Fc receptor-transfected fibroblasts. Science 1990; 249:641-646.

8. Matter K, Mellman I. Mechanisms of cell polarity: sorting and transport in epithelial cells. Current Opin Cell Biol 1994; 6:545-554.

9. Mostov K. Protein traffic in polarized epithelial cells: the polymeric immunoglobulin receptor as a model system. J Cell Science 1993; 17:21-26.

10. Lewis VA, Koch T, Plutner H et al. A complementary DNA clone for a macrophage-lymphocyte Fc receptor [published erratum appears in Nature 1986 Dec 18-31;324(6098):702]. Nature 1986; 324:372-375.

11. Ravetch JV, Luster AD, Weinshank R et al. Structural heterogeneity and functional domains of murine immunoglobulin G Fc receptors. Science 1986; 234:718-725.

12. Ravetch JV, Kinet JP. Fc receptors. Ann Rev Immunol 1991; 9:457-492.

13. Reth M. Antigen receptor tail clue. Nature 1989; 338:383.

14. Bonnerot C, Amigorena S. Murine low-affinity receptors for the Fc portion of IgG. Roles in cell activation and ligand internalization. Receptors Channels 1993; 1:73-79.

15. Bonnerot C, Daeron M. Biological activities of murine low affinity Fc receptors for IgG. Immunomethods 1994; 41-47.

16. Ravetch JV. Fc receptors: rubor redux. Cell 1994; 78:553-560.

17. Amigorena S, Bonnerot C, Drake JR et al. Cytoplasmic domain heterogeneity and functions of IgG Fc receptors in B lymphocytes. Science 1992; 256:1808-1812.

18. Muta T, Kurosaki T, Misulovin Z et al. A 13-amino-acid motif in the cytoplasmic domain of Fc gamma RIIB modulates B cell receptor signalling [published erratum appears in Nature 1994 May 26;369(6478):340]. Nature 1994; 368:70-73.

19. Daeron M, Latour S, Malbec O et al. The same tyrosine-based inhibition motif, in the intracytoplasmic domain of Fc gamma RIIB, regulates negatively BCR-, TCR-, and FcR-dependent cell activation. Immunity 1995; 3:635-646.

20. Miettinen HM, Rose JK, Mellman I. Fc receptor isoforms exhibit distinct abilities for coated pit localization as a result of cytoplasmic domain heterogeneity. Cell 1989; 58:317-327.

21. Miettinen HM, Matter K, Hunziker W et al. Fc receptor endocytosis is controlled by a cytoplasmic domain determinant that actively prevents coated pit localization. JCB 1992; 116:875-888.

22. Amigorena S, Salamero J, Davoust J et al. Tyrosine-containing motif that transduces cell activation signals also determines internalization and antigen presentation via type III receptors for IgG. Nature 1992; 358:337-341.

23. Odin JA, Edberg JC, Painter CJ et al. Regulation of phagocytosis and [Ca²⁺]i flux by distinct regions of an Fc receptor. Science 1991; 254:1785-1788.

24. Mao SY, Pfeiffer JR, Oliver JM et al. Effects of subunit mutation on the localization to coated pits and internalization of cross-linked IgE-receptor complexes. J Immunol 1993; 151:2760-2774.

25. Robinson MS. The role of clathrin, adaptors and dynamin in endocytosis. Curr Opin Cell Biol 1994; 6:538-544.

26. Robinson MS, Watts C, Zerial M. Membrane dynamics in endocytosis. Cell 1996; 84:13-21.

27. Trowbridge IS, Collawn J, Hopkins CR. Signal-dependent membrane protein trafficking in the endocytic pathway. Ann Rev Cell Biol 1994; 129-162.

28. Letourneur F, Klausner RD. A novel di-leucine motif and a tyrosine-based motif independently mediate lysosomal targeting and endocytosis of CD3 chains. Cell 1992; 69:1143-1157.

29. Daeron M, Malbec O, Latour S et al. Distinct intracytoplasmic sequences are required for endocytosis and phagocytosis via murine Fc gamma RII in mast cells. J Immunol 1993; 5:1393-1401.

30. Hunziker W, Fumey C. A di-leucine motif mediates endocytosis and basolateral sorting of macrophage IgG Fc receptors in MDCK cells. JCB 1994; 13:2963-2967.

31. Matter K, Yamamoto EM, Mellman I. Structural requirements and sequence motifs for polarized sorting and endocytosis of LDL and Fc receptors in MDCK cells. JCB 1994; 126:991-1004.

32. Budde P, Bewarder N, Weinrich V et al. Tyrosine-containing sequence motifs of the human immunoglobulin G receptors FcRIIb1 and FcRIIb2 essential for endocytosis and regulation of calcium flux in B cells. Eur J Cell Biol 1994; 269:30636-30644.

33. Romeo C, Kolanus W, Amiot M et al. Activation of immune system effector function by T cell or Fc receptor intracellular domains. Cold Spring Harbour Symposia on Quantitatie Biology, 1992; 57:117-125.

34. Felder S, Miller K, Moehren G et al. Kinase activity controls the sorting of the epidermal growth factor receptor within the multivesicular body. Cell 1990; 61:623-634.

35. Cresswell P. Assembly, transport, and function of MHC class II molecules. Ann Rev Immunol 1994; 12:259-293.

36. Germain RN, Margulies DH. The biochemistry and cell biology of antigen processing and presentation. Ann Rev Immunol [Review]. 1993; 11:403-450.

37. Lanzavecchia A. Receptor-mediated antigen uptake and its effect on antigen presentation to class II-restricted T lymphocytes. Ann Rev Immunol 1990; 8:773-793.

38. Lanzavecchia A. Antigen-specific interaction between T and B cells. Nature 1985; 314:537-539.

39. Manca F, Fenoglio D, Li PG et al. Effect of antigen/antibody ratio on macrophage uptake, processing, and presentation to T cells of antigen complexed with polyclonal antibodies. J Exp Med 1991; 173:37-48.

40. Esposito-Farese ME, Sautes C, de la Salle H et al. Membrane and soluble Fc gamma RII/III modulate the antigen-presenting capacity of murine dendritic epidermal Langerhans cells for IgG-complexed antigens. J Immunol 1995; 155:1725-1736.

41. Gosselin EJ, Wardwell K, Gosselin DR et al. Enhanced antigen presentation using human Fc gamma receptor (monocyte/macrophage)-specific immunogens. J Immunol 1992; 149:3477-3481.

42. Choquet D, Partiseti M, Amigorena S et al. Cross-linking of IgG receptors inhibits membrane immunoglobulin-stimulated calcium influx in B lymphocytes. JCB 1993; 121:355-363.

43. Griffin Jr FM, Griffin JA, Leider JE et al. Studies on the mechanism of phagocytosis. I. Requirements for circumferential attachment of particle-bound ligands to specific receptors on the macrophage plasma membrane. J Exp Med 1975; 142:1263-1282.

44. Mellman I, Koch T, Healey G et al. Structure and function of Fc receptors on macrophages and lymphocytes. J Cell Science 1988; 9:45-65.

45. Indik ZK, Park JG, Hunter S et al. The molecular dissection of Fc gamma receptor mediated phagocytosis. Blood 1995; 86:4389-4399.

46. Takai T, Li M, Sylvestre D et al. FcR gamma chain deletion results in pleiotrophic effector cell defects. Cell 1994; 76:519-529.

47. Axline SG, Reaven EP. Inhibition of phagocytosis and plasma membrane mobility of the cultivated macrophage by cytochalasin B. Role of subplasmalemmal microfilaments. J Exp Med 1974; 62:647-659.

48. Greenberg S, Burridge K, Silverstein SC. Colocalization of F-actin and talin during Fc receptor-mediated phagocytosis in mouse macrophages. J Exp Med 1990; 172:1853-1856.

49. Greenberg S, el Khoury J, di Virgilio F et al. Ca^{2+}-independent F-actin assembly and disassembly during Fc receptor-mediated phagocytosis in mouse macrophages. J Cell Biol 1991; 113:757-767.

50. Greenberg S, Chang P, Silverstein SC. Tyrosine phosphorylation of the gamma subunit of Fc gamma receptors, p72syk, and paxillin during Fc receptor-mediated phagocytosis in macrophages. J Exp Med 1994; 269:3897-3902.

51. Daeron M, Malbec O, Bonnerot C et al. Tyrosine-containing activation motif-dependent phagocytosis in mast cells. J Immunol 1994; 152:783-792.

52. Davis W, Harrison PT, Hutchinson MJ et al. Two distinct regions of FC gamma RI initiate separate signalling pathways involved in endocytosis and phagocytosis. EMBO J1995; 14:432-441.

53. Indik ZK, Park JG, Pan XQ et al. Induction of phagocytosis by a protein tyrosine kinase. Blood 1995; 85:1175-1180.

54. Indik ZK, Pan XQ, Huang MM et al. Insertion of cytoplasmic tyrosine sequences into the nonphagocytic receptor Fc gamma RIIB establishes phagocytic function. Blood 1994; 83:2072-2080.

55. Park JG, Schreiber AD. Determinants of the phagocytic signal mediated by the type IIIA Fc gamma receptor, Fc gamma RIIIA: sequence requirements and interaction with protein-tyrosine kinases. PNAS 1995; 92:7381-7385.

56. Greenberg S, Chang P, Wang DC et al. Clustered syk tyrosine kinase domains trigger phagocytosis. PNAS 1996; 93:1103-1107.

57. Aroeti B, Casanova J, Okamoto C et al. Polymeric immunoglobulin receptor. Int Res Cytology 1992; 137B:157-168.

58. Stuart SG, Simister NE, Clarkson SB et al. Human IgG Fc receptor (hFcRII; CD32) exists as multiple isoforms in macrophages, lymphocytes and IgG-transporting placental epithelium. EMBO J 1989; 8:3657-3666.

59. Casanova JE, Apodaca G, Mostov KE. An autonomous signal for basolateral sorting in the cytoplasmic domain of the polymeric immunoglobulin receptor. Cell 1991; 66:65-75.

60. Okamoto CT, Shia SP, Bird C et al. The cytoplasmic domain of the polymeric immunoglobulin receptor contains two internalization signals that are distinct from its basolateral sorting signal. JCB 1992; 267:9925-9932.

61. Casanova JE, Breitfeld PP, Ross SA et al. Phosphorylation of the polymeric immunoglobulin receptor required for its efficient transcytosis. Science 1990; 248:742-745.

62. Hirt RP, Hughes GJ, Frutiger S et al. Transcytosis of the polymeric Ig receptor requires phosphorylation of serine 664 in the absence but not the presence of dimeric IgA. Cell 1993; 74:245-255.

63. Hunziker W, Mellman I. Expression of macrophage-lymphocyte Fc receptors in Madin-Darby canine kidney cells: polarity and transcytosis differ for isoforms with or without coated pit localization domains. JCB 1989; 3291-3302.

SOLUBLE FC RECEPTORS

Catherine Sautès

In the early 1970s, soon after the discovery of Fc receptors on lymphocytes, molecules binding antigen-complexed IgG were detected in culture supernatants of mouse activated T cells and called IgG-binding factors (IgG-BF). These IgG-BF were hypothesized to be derived from membrane FcγR and to correspond to soluble forms of these receptors. Soluble forms of Fcγ receptors (sFcγR) were subsequently described in supernatants of cells of the immune system other than T cells and generalized to isotypes other than IgG. The cloning and identification of FcR genes allowed the molecular characterization of some of their soluble products and the demonstration that soluble FcR can be generated either by proteolytic cleavage of membrane FcR or by alternative splicing of TM-encoding exons of FcR genes.

The question of soluble FcR functions has been addressed by using natural or recombinant forms of soluble FcR and, more recently, in transgenic animals. A unifying concept emerges that soluble FcR are not only Ig-binding factors which interfere with Fc-dependent immune reactions but also molecules that interact with cell surface receptors and trigger or regulate immune functions through these receptors. We shall summarize and discuss our knowledge of the structures of soluble FcR, their mechanism of production and their biological functions. Their possible involvement in pathology will be discussed in chapter 7.

SOLUBLE Fcγ RECEPTORS (sFcγR)

SOLUBLE FcγR STRUCTURE, GENERATION AND LIGANDS

Mouse
The characteristics of the genes encoding the membrane forms of mouse low affinity FcγR and their products have been described in detail in chapter 2. Briefly, mouse FcγRII and the ligand-binding chain

Cell-Mediated Effects of Immunoglobulins, edited by Wolf Herman Fridman and Catherine Sautès. © 1997 R.G. Landes Company.

of mouse FcγRIII (α chain) are encoded by two distinct genes. The receptors share 95% amino acid sequence homology in their extracellular domains and differ in their intracytoplasmic regions. FcγRIIb1 and FcγRIIb2, the two membrane isoforms of FcγRII, differ by a 47 residues sequence which is present in the intracytoplasmic domain of b1. Whereas FcγRIIb1 is present on lymphocytes, FcγRIIb2 is expressed in macrophages and in cells. The highly homologous extracellular domains of FcγRII and of the α chain of FcγRIII cannot be distinguished by the currently available mAbs and have the same IgG-isotypic specificity.

Soluble forms of low affinity FcγR II and/or III are present in supernatants of activated T cells, T cell lines and T cell hybridomas, activated B cells, macrophage cell lines, and of epidermal Langerhans cells.[1-4] They can be detected in serum. In normal mice (Balb/c) grown in conventional animal care units, mean (+/-SD) levels (40 sera tested) are 337 +/- 189 ng/ml, as shown by a sandwich ELISA using the rat anti-FcγRII/III mAb 2.4G2 and a polyclonal rabbit anti-mouse sFcγRII.[5] Levels are low in germ-free animals,[3] increase after immunization, during parasitic or viral infections,[6] and in mice bearing tumors (see chapter 7).[7] Thus, serum levels of soluble forms of FcγR depend upon activation of the immune system. Soluble FcγR are also present in mouse colostrum (around 10 ng/ml) (C. Sautès unpublished data).

On lymphocytes and on NK cells, soluble FcγR are generated by cleavage of FcγRII or FcγRIII respectively, at a region close to the cell surface. This cleavage, which is made by enzymes as yet undefined, allows the release of soluble forms corresponding to the two extracellular domains of FcγRII or of the α chain of FcγRIII (Fig. 6.1). Whether FcγRII and FcγRIII proteolysis occur at similar sites has not been clarified. However, both polypeptides migrate at similar velocity by SDS-PAGE after deglycosylation, suggesting that close or identical cleavage sites may be involved for both receptors. Thus FcγRIIb1 expressed by activated B and T cells, and FcγRIII on NK cells, release in culture medium 38-40 kDa sFcγRII or III respectively, which give rise to a 18-19 kDa polypeptide after removal of N-linked carbohydrates.[8] The use of fibroblasts and FcγR-deficient B cell lines expressing high levels of recombinant FcγRIIb1 or b2 and externally labeled with ^{125}I allowed the demonstration that these sFcγR lack the intracytoplasmic sequences of the receptors, bind IgG and are generated by enzymatic cleavage of the cell membrane receptors, their production being concomitant with a decrease in cell surface expression.[8,9]

On macrophages and on Langerhans cells, which express FcγRIIb2 and FcγRIII, sFcγR are produced not only by enzymatic cleavage of FcγRIIb2 and FcγRIII, as described above, but also by alternative splicing. Thus, a soluble form called FcγRIIb3, corresponding to FcγRIIb2 without transmembrane region sequences, has been detected in culture supernatants of J774 and P388D1 macrophage cell lines, and of epidermal Langerhans cells, by western blots and by ELISA using antibodies directed

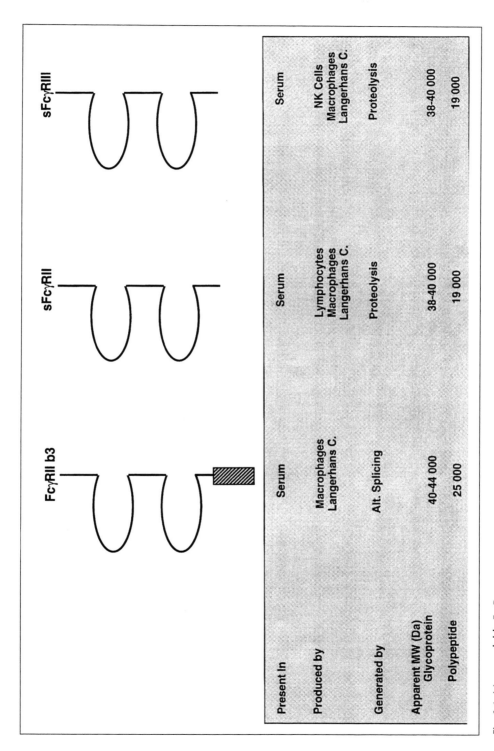

	FcγRII b3	sFcγRII	sFcγRIII
Present In	Serum	Serum	Serum
Produced by	Macrophages Langerhans C.	Lymphocytes Macrophages Langerhans C.	NK Cells Macrophages Langerhans C.
Generated by	Alt. Splicing	Proteolysis	Proteolysis
Apparent MW (Da) Glycoprotein	40-44 000	38-40 000	38-40 000
Polypeptide	25 000	19 000	19 000

Fig. 6.1. Mouse soluble FcγR.

towards the COOH terminal end of FcγRII.[5,9,10] It was also identified in these cell types and in thioglycollate-elicidated macrophages by PCR amplification.[10] FcγRIIb3 isolated from P388D1 culture supernatants migrates as a 45-50 kDa glycoprotein and gives rise, after deglycosylation by N-glycanase, to a 25 kDa polypeptide.[10] Indeed this polypeptide contains the ectodomain and the intracytoplasmic domains of FcγRIIb2, as shown by peptide mapping and sequencing of PCR product.

Biochemical and immunochemical analysis of sFcγR isolated from mouse serum appears as 40-50 kDa glycoproteins that give two FcγR polypeptides after deglycosylation: a 25 kDa one corresponding to FcγRIIb3 and a 19 kDa one.[10] Whether the latter corresponds to FcγRII and/or FcγRIII ectodomains is unclear. In mice bearing lymphoid tumors that have a defect in the FcγRII gene, FcγRIIb3 levels increase, suggesting indeed the existence of regulatory mechanisms controlling FcγRIIb3 expression in macrophages or Langerhans cells.[7]

Recombinant forms of sFcγRII and FcγRIIb3 have been produced in eukaryotic cell lines by transfection with a cDNA encoding FcγRIIb1 mutated by insertion of a stop codon in the extracellular region coding sequence, at Lys175, a position near the transmembrane region, or with a cDNA encoding FcγRIIb3 respectively.[5,11,12] Several cell lines have been obtained in mouse fibroblastic L cells and in baby hamster kidney cells. The co-transfection of BHK cells with two selectable genes (the dihydrofolate and the neomycin resistance genes), followed by the culture of transfectants on hollow fibers in a perfusion cell culture system, yielded quantities of sFcγRII or FcγRIIb3 ranging from 100-500 mg per bioreactor. Soluble FcγR was purified from these culture fluids by three steps chromatography, which include ion exchange, affinity chromatography on rabbit IgG-coupled sepharose, and gel exclusion chromatography. When produced in BHK cells, soluble FcγRII corresponding to FcγRII ectodomain has an apparent Mr between 32 and 34,000 and a pI between 7 and 9, and FcγRIIb3 an apparent Mr between 47 and 56,000 and a pI between 4.8 and 6.6. After removal of N-linked polysaccharides, the Mr decrease to 19,000 and 26,000 respectively. The recombinant polypeptides have the same migration velocity in SDS-PAGE gels as the natural ones. The distinct molecular sizes between natural and recombinant glycoproteins reflect cell type-specific glycosylation processes.

Soluble FcγR bind the same isotypes than the membrane receptors and react with the same mAb, 2.4G2.[8,12] Thus sFcγRII obtained by cleavage of cell surface FcγRIIb1 or b2 bind insolubilized mouse IgG1, IgG2a, IgG2b but not IgG3 and FcγRIIb3 from mouse serum can be purified by affinity chromatography on IgG1, IgG2a or IgG2b but not on IgG3 or F(ab')2 fragments of IgG1 coupled to sepharose. These observations were confirmed by using the recombinant versions of proteolytically-cleaved sFcγRII and of FcγRIIb3 made in fibroblastic L cells[8,12] and in BHK cells[5] and both soluble FcR were found to bind rabbit

IgG that could be used for purification under insolubilized form.[5,11] Indeed the affinity is low, binding of these sFcγR to IgG requires low ionic strength buffers and elution can be achieved using mild acidic solutions.[5,8,11] Both types of sFcγR are relatively stable and can easily be kept frozen as shown by SDS-PAGE, which is not the case for all human sFcγRII isoforms, as discussed further.

The interaction between sFcγRII and human IgG1 has been studied by ^{1}H NMR using labeled Fc fragment and peptides of human IgG1 and recombinant sFcγRII. The Fc fragments were shown to interact with the receptor through multiple topographically distinct sites including Leu234,[13] extending previous mapping studies which implicated the C_H2 domain as the primary binding site on IgG for mouse FcγRII.[14,15]

In man, soluble FcR were found to have other ligands than IgG (see further) and to bind to cell surface receptors. The existence of cellular ligands for soluble FcγR was extended to mouse. Mouse sFcγRII was found to bind to the surface of cells of the immune system, as shown by cytofluorimetry using biotinylated recombinant sFcγRII followed by streptavidin phycoerythrin. A significant proportion of spleen, lymph node and bone marrow cells, (J. Galon, unpublished data) and a minor proportion of fetal thymic cells with a Pgp-1-, Thy-1- and surface immunoglobulin phenotype[16] bind sFcγRII. The type of interaction and of cell surface receptors involved remain to be solved.

Human

Soluble FcγRI (sFcγRI)

Of the three genes that encode a high affinity FcγR, only FcγRIA encode a membrane receptor and FcγRIB and FcγRIC contain stop codons in the exon encoding the third extracellular domain. Transcripts FcγRIb and FcγRIc have been detected. However no soluble forms of FcγRI have been described at the protein level.

Soluble FcγRII (sFcγRII)

As described in chapter 2, three FcγRII genes, FcγRIIA, FcγRIIB, FcγRIIC that encode a total of five distinct cell surface receptors with strong homologies in the extracellular and transmembrane regions (85% overall identity), exist in man. Human FcγRII are expressed on all cells of the immune system with the exception of some T cells and NK cells. They are also present on platelets and on non-immune cells such as trophoblasts or some endothelial cells. Whereas FcγRIIB products are found in B cells and in myelomonocytic cells, FcγRIIa1 is widely expressed on various cell types.

Soluble forms of FcγRII or CD32 (sFcγRII or sCD32) are present in supernatants of activated B cells,[17] epidermal Langerhans cells,[18] of the megakaryocytic cell line DAMI,[19] and in those of activated platelets.[20] Human biological fluids such as serum or unstimulated saliva contain significant levels of sFcγRII, as shown by immunodot assays and ELISA.[5,21,22]

In man, as in mouse, sFcγRII can be formed by proteolytic cleavage of membrane FcγRII (Fig. 6.2). This was shown to be the case for FcγRIIB products expressed by anti-IgM or phorbol 12 myristate 13-acetate activated B cells[17,23] and for FcγRIIa1 present on epidermal Langerhans cells.[18] However, human products are unstable, in contrast to the murine equivalents. The ability of activated—but not resting—B cells to release sFcγRII correlates with the expression of early activation markers and with the appearance of a trypsin-like serine protease activity in the same cells. Addition of serine protease inhibitors prevents the release, suggesting that release is a consequence of FcγRII proteolytic cleavage.[17,23] The fragments have an apparent molecular mass of 33,000 and 14,000-18,000 under non reducing conditions. Soluble FcγRII has been also detected in culture supernatants of CHO cells expressing recombinant FcγRIIa1. It seems to be also very unstable and has not been biochemically characterized yet.[18]

A FcγRIIA mRNA lacking the TM domain sequence has been detected by PCR analyses in human platelets,[20] the megakaryocyte cell line HEL,[24] the erythromyeloid cell line K562,[25] the histiocytic cell line U937[5,20] and human epidermal Langerhans cells.[18] FcγRIIa2, its product, is a glycoprotein reacting with IV.3, a mAb directed against the extracellular region of FcγRIIa1 and with antibodies against the intracytoplasmic region of FcγRIIa1. The secreted FcγRIIa2 polypeptide has been detected in culture supernatants of cells expressing FcγRIIa1. Its production is maximal during the first 3 hours of incubation at 37°C of human Langerhans cells at 37°C. FcγRIIa2 glycoproteins have apparent Mrs between 35,000 and 39,000, depending on the cell type. When expressed in CHO cells, the apparent molecular mass is between 43,000 and 47,000. Of note, FcγRIIa2 polypeptide exhibits limited size heterogeneity in various cell types, as shown by SDS-PAGE after removal of N-linked polysaccharides. Forms with apparent Mr of 32,000 have been described in supernatants from epidermal Langerhans cells after deglycosylation, whereas the megakaryocytic cell line DAMI and thrombin-activated platelets release FcγRIIa2 that give a 24 kDa polypeptide after removal of N-linked sugars.[19] This FcγRIIa2 has lost a 10 kDa COOH terminal sequence corresponding to the COOH-terminal part of the molecule, as shown by western blotting. Noteworthy, COOH-terminal degradation was also observed in recombinant FcγRIIa2 produced in CHO cells, both at the level of the glycoprotein and of the polypeptide.[18]

Cytokines can regulate FcγRIIa2 expression. TNF-α increases FcγRIIa2 expression in Langerhans cells, as shown at the mRNA level. This mRNA predominates over the membrane form encoding mRNA after up to 24 hours of culture.[26,27]

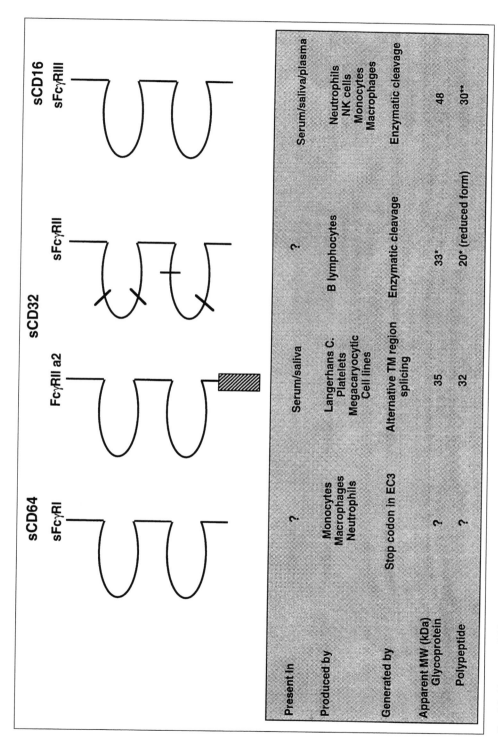

	sCD64 sFcγRI	sCD32 FcγRII a2	sCD32 sFcγRII	sCD16 sFcγRIII
Present in	?	Serum/saliva	?	Serum/saliva/plasma
Produced by	Monocytes Macrophages Neutrophils	Langerhans C. Platelets Megacaryocytic Cell lines	B lymphocytes	Neutrophils NK cells Monocytes Macrophages
Generated by	Stop codon in EC3	Alternative TM region splicing	Enzymatic cleavage	Enzymatic cleavage
Apparent MW (kDa) Glycoprotein	?	35	33*	48
Polypeptide	?	32	20* (reduced form)	30**

Fig. 6.2. Human soluble FcγR : (*) sensitive to proteolysis, (**) NA1 and NA2 forms slightly differ.

Soluble FcγRII that circulate in human serum correspond to shed FcγRII and to FcγRIIa2. As shown by a sandwich ELISA using IV.3 mAb and a rabbit polyclonal IgG directed against the intracellular region of FcγRIIa1, the FcγRIIa2 levels range from 0 to 30 ng/ml, the mean value being 11.9 ± 6.55 ng/ml and the median one 10.6 ng/ml in a study performed on sera from 51 healthy donors.[22]

The IgG-binding capacities of FcγRIIa2 have been demonstrated by inhibition of rosette formation between FcγRIIa1 positive cells (K562) and sheep red blood cells (SRBC) sensitized with different IgG mAbs directed to SRBC, and by binding to insolubilized IgG, followed by elution and SDS-PAGE analysis. Studies have been performed with recombinant FcγRIIa2 cloned from Langerhans cells from a "high responder" donor and expressed in the dihydrofolate reductase-deficient CHO-DG44 cell line. Like FcγRIIa1, recombinant FcγRIIa2 binds indeed human IgG1, IgG3, and to a lesser extent IgG4, but does not bind IgG2. Specificity for mouse IgG is IgG1>IgG2b>IgG2a>>>IgG3.[18]

Soluble FcγRIII (sFcγRIII)

Two genes encode the low affinity type 3 receptors in man, hFcγRIIIA and hFcγRIIIB. Their expression is tissue-specific, FcγRIIIA being expressed in NK cells and monocytes, and FcγRIIIB in neutrophils. The FcγRIIIA product is a transmembrane glycoprotein with two disulfide-linked extracellular domains, non covalently bound to a dimer of signal transducing chains. The highly homologous FcγRIIIB is synthesized as a precursor with a hydrophobic transmembrane-like region and a short intracytoplasmic tail, cleaved and anchored by a glyco-phosphatidyl inositol link to the cell membrane.

Soluble FcγRIII (Fig. 6.3) have been detected in serum and in saliva. Few µg/ml ($1.25 + 0.28$ µg/ml, 30 donors tested) circulate in serum from healthy donors as shown by a sandwich ELISA using 3.G8 mAb and a rabbit anti-sFcγRIII polyclonal IgG.[5] The plasma clearance is 0.7 ml/min and the half-life about 1.8 day.[28] Since FcγRIIIa and FcγRIIIb are very homologous in their extracellular domains, the question of the cellular origin of serum sFcγRIII has been addressed. The vast majority is produced by neutrophils and corresponds to shed FcγRIIIb because patients in whom the expression of FcγRIIIB is low because of a defect in pi-linkage of cell surface proteins, such as in paroxysmal nocturnal hemoglobinuria, (PNH)[29]or patients deficient in the FcγRIIIB gene[30] have low or undetectable levels of sFcγRIII. The plasma concentrations depend mainly on the production of neutrophils in the bone marrow, as shown in patients with high dose chemotherapy followed by G-CSF administration. In these patients, sFcγRIII plasma levels drop and increase in parallel to these treatments. In contrast, the concentration of sFcγRIII is not correlated with the number of circulating neutrophils and is unaffected by shifts between the storage, circulating and marginating pools.[28] Thus sFcγRIII in plasma is a measure of the total body neutrophil mass and thereby of neutrophil defense. FcγRIII has two allelic forms NA1 and NA2. NA phenotype-dependent differences in the plasma concentration of sFcγRIII may exist, that seem to be related to subtle differences in membrane expression of FcγRIIIb, levels being slightly lower in NA1-positive than in heterozygous and NA2 -positive donors.[31]

Soluble FcγRIIIa was isolated from the plasma of FcγRIIIB negative donors with a complete absence of both FcγRIIIB genes.[32] SDS-PAGE analysis showed that its mobility was identical to that of sFcγRIIIa shed by NK cells (45 to 55 kDa) and not to that of the slightly larger sFcγRIIIa shed by monocytes (45 to 72 kDa). Patients suffering from an NK-cell lymphocytosis showed high levels of sFcγRIIIa in their plasma, that had the same size as sFcγRIIIa shed by NK cells in vitro. Nevertheless, sFcγRIIIa derived from NK cells was present in very low amounts as compared to sFcγRIIIb in plasma from control patients.

Neutrophils lose their membrane FcγRIIIb when activated by PMA, or kept in culture for 18 to 24 hours. Soluble FcγRIII are produced in parallel. The PMA-dependent cleavage of membrane FcγRIII involves metalloprotease and serine protease activities inhibitable by 1,10-phenantroline and TLCK (Na-p-tosyl-L-lysine chloromethyl ketone), respectively.[33,34] The metalloprotease seem to be dependent on a heavy metal cation because 1,10-phenantroline blocks the cleavage more efficiently than EDTA. It remains unclear whether two enzymes are necessary or whether a serine protease, the activity of which is

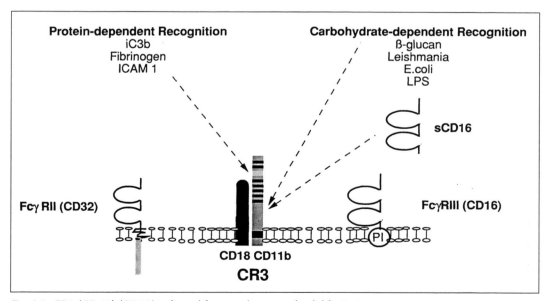

Fig. 6.3. CR3 (CD11b/CD18), a ligand for membrane and soluble FcγRIII.

dependent on a heavy metal cation, might be involved. Collagenase or gelatinase, both metalloproteases known to be expressed by granulocytes in a latent inactive form and activated by serine proteases after cell stimulation are candidates for these enzymes. Elastase treatment of freshly isolated neutrophils reduces FcγRIII expression by 90%.[35] Membrane metalloproteases of the same family as those involved in the cleavage of the p80-IL-6 receptor or the p60-TNF receptor might also participate.[36]

A far from negligible amount of FcγRIII is present within neutrophils, sequestered in intracellular compartments. Thus FcγRIII have been detected in secondary specific granules and in intracellular alkaline phosphatase-containing membranes.[35] It is noteworthy that incubation of neutrophils with the chemoattractant FMLP does not modify cell surface expression of FcγRIII since it causes a rapid translocation of intracellular FcγRIII to the PMN surface that is roughly balanced by the concomittant FMLP-induced shedding of this receptor. Thus, the surface expression of FcγRIII reflects a balance between synthesis, shedding, and secretion from intracellular stores.

Several studies have indicated that neutrophil death occurs by apoptosis, leading to macrophage recognition and phagoytosis that limits tissue injury and promote resolution of inflammation. During in vitro culture, neutrophils die by apoptosis, as shown by dramatic morphologic alterations (nuclear condensation, vacuolation) and lose their membrane CD16 in parallel.[37] The apoptotic population has acquired the property of binding annexin-V, a calcium-dependent, phospholipid-binding protein with high affinity for phosphatidyl-serine.[38] Addition of IFN-γ, GM-CSF or G-CSF to the culture diminishes the FcγRIII membrane loss. The proteolytic enzymes involved FcγRIII release during apoptosis are still unknown.

On NK cells, FcγRIIIa is also released by proteolysis which occurs spontaneously or upon stimulation with PMA.[39] The polypeptide (29 kDa) is smaller than the membrane-bound molecule (34-36 kDa). A metalloprotease may be involved, as shown by the inhibitory effect of 1,10-phenantroline on shedding. Several serine-protease inhibitors have no effect, and iodoacetamide slightly enhances the release.

A stop-linker was introduced in the cDNA encoding the NA2 form of FcγRIIIb, resulting in the production of a 194 amino-acid-long soluble form containing six accessory amino acid residues.[40] Using recombinant sFcγRIII produced in fibroblastic L cells and in BHK cells, sFcγRIII was shown to bind insolubilized human IgG1, IgG3, but not human IgG2 and IgG4 and to react with 3.G8, BW 209,[40] (J. Galon, personal communication) and B73. Thus sFcγRIII exhibit the same specificity as the membrane receptor (chapter 2). A soluble FcγRIIIb produced in *E. coli* has the same IgG-binding profile suggesting that sugars do not play an important role in the isotypic specificity of FcγRIII (J. Galon in preparation).

Of note, recent studies using natural sFcγRIII purified from human serum and eukaryotic recombinant sFcγRIIIb have indicated that sFcγRIII binds to the cell membrane via ligands other than IgG. Soluble FcγRIII was biotinylated and its binding to human blood cells and cell lines was investigated by cytofluorimetry using streptavidin phycoerythrin conjugate. These studies have shown that sFcγRIII bind to CR3, complement receptor 3, and CR4, complement receptor 4-positive leukocytes and cell lines (such as the monocytic cell lines THP1 and U937), and to transfected cells expressing the recombinant form of the receptors. All monocytes and neutrophils bind sFcγRIII and minor proportions of B and T cells as well. The labeling is inhibited by some anti-CR3 and CR4 antibodies, and the upregulation of CR3 and CR4 triggered by PMA or LPS-activation of neutrophils leads to an increased fixation of sFcγRIII.[41]

The complement receptors CR3 and CR4 are β2 integrin heterodimers, named CD11b/CD18 and CD11c/CD18 respectively, that bind a series of proteic and glucidic ligands. The proteic ligands of CR3 include fibrinogen, factor X, ICAM1, in addition to complement component iC3b. Polysaccharides like LPS and zymosan also interact with CR3. A series of epitope mapping studies have shown that whereas iC3b, fibrinogen, and ICAM1 bind to the I domain of the α chain of CR3 (CD11b), the polysaccharide-binding sites map in the C-terminal part of the extracellular region of CD11b.

The lectin-like part of CR3 plays an important role in the binding of sFcγRIII since the interaction between sFcγRIII and CR3 is inhibited by anti-CD11b mAbs that map in this site (OKM1, VIM12) but not by these that map to the I domain of CD11b (Mo1, 44 and OKM10) and by sugars such as N-acetyl D glucosamine, α or β-methyl D mannoside, α or β-methyl D glucoside and zymosan that also bind to it. On neutrophils, the GPI-linked FcγRIIIb interacts and cooperates with CR3, as shown by co-capping experiments,[42] lateral diffusion coefficient measurements[43] and biological studies,[44] and the lectin-like site of CD11b is involved in the interaction. Thus sFcγRIII bind to CR3 in a similar manner than the GPI-linked FcγRIII, i.e. via the lectin-like part of CD11b (Fig. 6.3).[41] The mode of interaction of sFcγRIII with CR4 has not been investigated yet.

SOLUBLE FCγR FUNCTIONS

IgG-binding functions

As described above, sFcγR produced by cleavage of FcγR or by alternative splicing contain the extracellular region of the receptors. They bind to IgG-containing immune complexes and have the same isotypic specificity than the entire corresponding receptors (details given chapter 2).

One of the main functions of sFcγR is indeed to interact with IgG immune complexes and, by competition, to block Fc-dependent immune reactions. Thus a recombinant human sFcγRII was found to block IgG-dependent reverse Arthus reaction in rats and human FcγRIIa2 was shown to inhibit anti-CD9-mediated platelet aggregation.[19] Presentation of IgG immune complexes by APC occurs via the low affinity FcγR which internalize the immune complexes and present Ag to sensitized T cells (chapter 5). The role of sFcγR in this process was recently investigated in a murine model. Mouse epidermal Langerhans cells express FcγRIIb2, FcγRII b3 and FcγRIII and internalize FcγR by receptor-mediated endocytosis. FcγR-mediated Ag internalization improves by 300-fold their Ag-presenting capacity to T cells. The preincubation of IgG-complexed Ag with sFcγRII or with FcγRIIb3 led to a dose-dependent decrease in the Ag-presenting capacity of Langerhans cells to clones of T helper cells.[4] The essential function of Langerhans cells is to capture epicutaneously applied Ag. Since Langerhans cells secrete sFcγR, one may speculate that sFcγR may protect, up to a

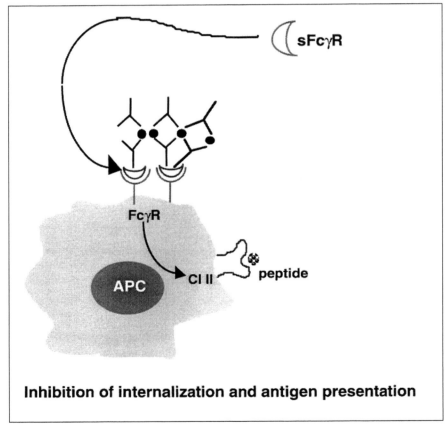

Inhibition of internalization and antigen presentation

Fig. 6.4. Murine soluble FcγRII inhibits internalization of IgG immune complexes by Langerhans cells and subsequent increase in antigen presentation on class II molecules.

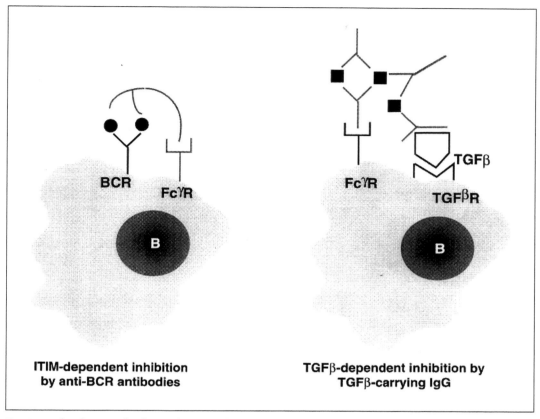

**ITIM-dependent inhibition
by anti-BCR antibodies**

**TGFβ-dependent inhibition by
TGFβ-carrying IgG**

Fig. 6.5. Role of IgG antibodies and TGFβ in the inhibition of B cell response. On the left, the crosslinking of BCR and FcγRIIb1 on B cells by IgG antibodies inhibits B cell activation towards Ag recognized by B cells. On the right, IgG immune complexes carry TGF-β to B cells and inhibit phase S entry.

certain extent, Langerhans cells from processing immune complexes of endogenous origin (Fig. 6.4).

Targeting of Ag to FcγR enhances the efficiency of presentation by 100 to 1000 times compared with that obtained with soluble Ag. Antibodies allow recognition of minute amounts of Ag that cannot be taken up efficiently by accessory cells. Thus, by decreasing Ag-IgG presentation, sFcγR may decrease subsequent immune reaction during ongoing responses and participate in the regulation of the response.

One of the questions is whether sFcγR are involved in immune reactions against tumors. Destruction of tumor cells by cytotoxic cells involves different mechanisms, one of which is the antibody-dependent cell-mediated cytotoxicity (ADCC) directed towards IgG-sensitized tumor cells. In a human model, using peripheral blood mononuclear cells as effector cells, and, as targets, cells from a human neuroblastoma line, LAN1, sensitized with a chimeric antibody having a Fab region directed against the diasialoganglioside GD2 expressed by LAN1 and an Fc region derived from a human monoclonal IgG1, it was shown that the recombinant FcγRIIa2 inhibits target cell lysis.[5] Thus, by decreasing ADCC, and by decreasing the host immune response, sFcγR may play an inhibitory role in immune reactions against tumors. In mice, serum sFcγR levels were found to increase during tumor growth.[7] For instance, mice bearing Ig-secreting tumors have up to ten times more circulating sFcγR than control animals, sFcγR being produced by the host during tumor response. This sFcγR increase may amplify host immunosuppression that occurs in tumor-bearing animals.

Finally, like their membrane forms, soluble FcγR most likely play an important role in non-organ specific autoimmunity by interfering with the binding of IgG-immune complexes to effector cells (see chapter 7).

Via their Fc portion, antibodies trigger various functions including internalization or phagocytosis, cytokine release, superoxide production or ADCC. By interfering with the binding of IgG-immune complexes to FcγR, sFcγR downmodulate these reactions and can be viewed as negative regulators of the immune system.

Other functions

In the mid-seventies, IgG-binding factors (IgG-BF) that were produced by FcγR positive T cells and hypothesized to be derived from membrane FcγR, were found to inhibit in vitro antibody production to T-dependent and T-independent antigens.[45,46] On B cell hybridomas, IgG-BF was found to decrease IgG secretion as soon as two hours after addition, and to exert a strong growth inhibitory effect.[47] Mitogen-induced human peripheral B cell differentiation is also suppressed by addition of lymphocyte-derived IgG-BF.[48] Recombinant mouse sFcγRII were later on also found to inhibit IgM and IgG antibody production to SRBC, in vitro[12] and in vivo[9] and the anti-IgM or LPS-induced proliferation of mouse B cells.[5,49,50]

Several lines of evidence have also indicated that sFcγR may be of importance for maintenance of local immunosuppression during pregnancy.[51] Human trophoblasts which express FcγRII produce molecules that suppress lymphocyte proliferation in allogeneic cultures and bind to the Fc portion of IgG. Retroplacental serum contain sFcγR that suppresses phytohemagglutinin-stimulated lymphoproliferation. Thus by their anti-proliferative properties, it has been hypothesized that sFcγR may participate in the escape from allograft rejection that occurs during pregnancy.

Cells of the immune system produce endogeneous immunoregulatory molecules such as transforming growth factor-β, TGF-β. In addition to its initially described activity as transforming factor, TGF-β exerts pleiotropic activities regulating many inflammatory processes and having a high anti-proliferative activity on many cell types. Since cells secreting

IgG-BF (activated T cells, B cells, monocytes) or recombinant sFcγR (produced in BHK cells) produce TGF-β, experiments were undertaken to investigate the respective roles of TGF-β and mouse sFcγRII/III in the suppressive activity of IgG-BF on mouse B cells. Based on kinetic studies, the neutralizing effect of anti-TGF-β antibodies and gel exclusion chromatography, results showed that TGF-β, but not sFcγR, was responsible for the suppressive activity of IgG-BF. Moreover, TGF-β was found to bind to IgG via its Fc portion, supporting the concept that TGF-β is an IgG-BF.[49]

TGF-β is synthesized and secreted as a latent molecule which does not bind to the TGF-β receptors and that can be activated into a biologically active molecule. It has been shown that immune complexes of IgG antibodies isolated from serum carry TGF-β which suppresses cytotoxic T cell responses to unrelated alloantigens.[52] Thus TGF-β delivered by IgG immune complexes may mediate a negative regulatory effect. For instance, on B cells, TGFβ may amplify the inhibitory signal triggered by the crosslinking of IgG antibodies to the B cell receptor (Fig. 6.5).

The recent identification of cell surface receptors for sFcγR leads to the concept that sFcγR may not be only competitors for the binding of IgG to membrane FcγR, but also may trigger immune reactions and behave like cytokines. In man, sFcγRIII was found to bind to

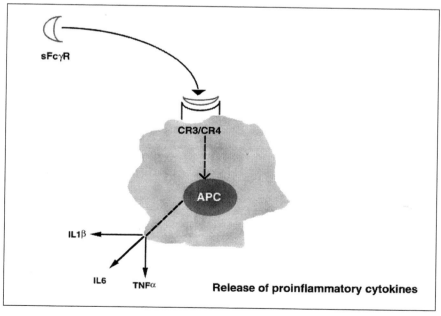

Fig. 6.6. Soluble CD16 induces production of proinflammatory cytokines by human blood monocytes through interaction with CR3 (CD11b/CD18) or CR4 (CD11c/CD18).

CR3 and CR4.[41] The interaction with CR3 involves the lectin-like part of the α chain of CD11b. On monocytes, the engagement of CR3 by specific antibodies triggers activation signals such as intracellular Ca^{2+} increase and leads to TNF-α secretion. Recent experiments showed that endotoxin-free sFcγRIII induces the CR3-dependent release of proinflammatory cytokines such as IL-6 and IL-8 by monocytes, a significant production of these cytokines requiring a minimal dose of 0.1 µg/ml.[41] The IL8-release by neutrophils stimulated by zymosan is dependent on a CR3-signaling pathway. Incubation of neutrophils with sFcγRIII increased their IL-8 production, 24 to 48 hours after stimulus addition.[41] Thus, sFcγRIII trigger activation signals and pro-inflammatory cytokine release via its binding to CR3 (Fig. 6.6). These observations are reminiscent of those made with sCD23, the soluble form of the low affinity FcεR, as discussed later. The identification of these new ligands for sFcγRIII opens the possibility that sFcγRIII may play a role in inflammatory reactions. Soluble FcγRIII is produced by neutrophils. Upon binding to endothelial cells,

Fig. 6.7. Membrane and soluble CD23 (Schematic representation).

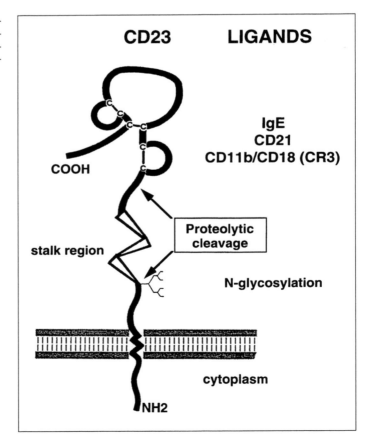

neutrophils are activated and may release sFcγRIII that could trigger and amplify the inflammatory reaction.

Dendritic cells play a key role in immune response. In the skin, the immature dendritic cells are powerful antigen-presenting cells. They internalize, process and present Ag to T cells. They mature and migrate to secondary lymphoid organs where they deliver costimulatory signals for T cell activation: they produce IL-12 and express the costimulatory molecules, B7 and CD40. Immature dendritic cells can be obtained by culture of blood monocytes with IL-4 and GM-CSF. Recent experiments have shown that the incubation of monocyte-derived dendritic cells with multivalent sFcγRIII leads to the modification of cell surface markers and to the production of high levels of IL-12 and GM-CSF suggesting that sFcγRIII may lead to Th1 type response (Galon et al in preparation) (Fig. 6.7). These findings open a new array of biological functions for sFcγR in immune defense.

SOLUBLE FcεR

SOLUBLE FcεR STRUCTURE, GENERATION AND LIGANDS

Soluble FcεRI

The high affinity FcεR is composed of three chains, a ligand-binding chain which belongs to the super Ig family, α, a β chain with five transmembrane regions and a γ chain, similar to the γ chain of other FcR. Cells involved in allergic reactions such as mastocytes and basophils express FcεRI. No information is available about their capacity to produce sFcεRI or whether sFcεRI is detectable in serum. Nevertheless, since FcεRI plays a decisive role in the triggering of primary mediators of allergy, recombinant sFcεRI have been produced by a number of groups to tentatively manipulate allergic reactions. Inhibition of recurrent allergic reactions have been described in a mouse model of type I allergy.[53] Passive cutaneous anaphylaxis response in rats was also depressed after soluble FcεRI injection.[54]

Soluble FcεRII

The characteristics of FcεRII (CD23) have been described in details in chapter 2. Briefly, FcεRII is a 45 kDa type II transmembrane glycoprotein present on mature B cells, APC such as monocytes, macrophages, Langerhans cells and, at lower levels, on activated T cells. Its expression level is regulated by cytokines such as IL-4, IL-13, IFNγ, by mediators such as LTB4, PAF, 1,25 dihydroxyvitamin D3, and by the triggering of other cell surface receptors such as CD40, CD72, sIg, and CD20.

Soluble forms of FcεRII have been described in the mid-seventies in supernatants of T lymphocytes incubated with IgE and called IgE-binding factors (IgE-BF). Depending on their glycosylation, IgE-BF may suppress or enhance the synthesis of IgE by activated IgE-bearing B cells.[55,56] In mouse, the genes encoding these factors have been cloned and were found to be a variant member of the endogeneous retroviral gene family related to mouse intracisternal A particles.[57,58] These IgE-BF have no structural homologies with the soluble forms of the animal lectin family member FcεRII or CD23 that was subsequently isolated and cloned from human B cells or B cell supernatants[59,60] and that will be reviewed herein.

When isolated from supernatants of human B cell line, sCD23 exhibit size micro-heterogeneity with molecular weights ranging from 37, 33, 25-27, and 16 kDa.[61] The 37 and 33 kDa fragments are short-lived intermediates that are cleaved into 25-27 kDa molecules. This breakdown is inhibited by iodoacetamide. Protein sequencing has shown that the 37 kDa fragment is cleaved at amino acid 82 of the CD23 molecule, the 33 kDa at 102, the 25-27 kDa at positions 148 to 150. In mouse, sFcεRII is also heterogeneous in size and contains 38, 35, 28 and 25 kDa components (Fig. 6.8). While similar sized fragments are generated in mouse and in man, there are no obvious homologies between the cleavage sites in both species. In man, the protease responsible for this breakdown is associated with FcεRII, if not FcεRII itself.[62] Of note, mechanisms other than proteolytic cleavage may be involved in sCD23 production. Transcripts lacking the entire third exon which encodes the transmembrane region and part of the cytoplasmic tail have been found in human B cell line RPMI8866.[63]

Recent studies in man and in mouse have indicated that what is termed the stalk region of the CD23 molecule, i.e. the region between the COOH-terminal lectin-homology domain and the membrane surface, contains a heptad periodicity that is characteristic of an α-helical coiled-coil.[64] The stalk region participates in the formation of CD23 oligomers on the cell surface. Analysis of the cleavage sites in man and in mouse demonstrated that the proteolytic sites involved in the release of the 25 kDa fragment are located within this region. In mice, mutant FcεRII with a disruption in the α-helical coiled-coil have a monomeric interaction with IgE, likewise with 25 kDa sFcεRII which lack this region. In contrast, as shown in man, the murine 37 and 33 kDa fragments have significant lengths of potential coiled-coil. As with membrane CD23, they can form dimers or trimers. Additional interactions between the lectin heads result in the formation of hexamers in solution. These observations may have functional consequences since a trimeric 37 or 33 kDa sCD23 would function more efficiently than a 25 kDa one, due to avidity considerations, with regard to cytokine-inducing activities or to blockade of IgE binding to CD23.

The production of sCD23 is dependent upon environmental factors. Ligands such as IgE and some anti-CD23 reduce the rate of CD23 cleavage.[65] Glucocorticoids decrease sCD23 formation. In contrast, engagement of CD20 on B cells stimulates the cleavage.[66] Addition of

IL-4 to human B cell cultures induces the production of sCD23.[67] This effect is up regulated by TNF-α.[68]

Soluble CD23 circulates in serum and its levels are enhanced in allergic and in inflammatory reactions, as described in chapter 7.

Functions of sFcεRII

Both murine and human sFcεRII interact poorly with IgE. As compared to man, the affinity of the mouse product is reduced by at least an order of magnitude. As described above, avidity of the various sFcεRII fragments for IgE varies according to polypeptide length.

In humans, native sCD23 was found to augment the production of IgE by normal PBMC stimulated with suboptimal concentrations of IL-4.[61,69] More recently, the cytokine-like activities of sCD23 were extensively studied using a 25 kDa recombinant sCD23. In the presence of IL-1, sCD23 induces differentiation of myeloid and T cell precursors,[70-72] promotes the survival of germinal center B cells and also drives them toward a plasmacytoid pathway of differentiation.[73] These activities have led to the identification of new ligands for full length CD23, such as CD21,[74] CD11b/CD18 and CD11c/CD18.[75] The lectin-like nature of CD23 most likely is decisive for these interactions, CD23 being incapable of binding to non-N-glycosylated forms of these molecules. The CD23/CD21 interactions may facilitate T-B cooperation for IgE production in vitro and CD23/CD11b/CD18 interactions may be involved in inflammatory responses. Whether sCD23 binds as well to these receptors is unknown. The fact that anti-CD21 antibodies exhibit activities like sCD23 with respect to inhibition of germinal center apoptosis suggests that it may be involved in this function. In a recent study, sCD23 was found to activate monocytes to contribute to the antigen-independent stimulation of resting T cells.[76] Soluble CD23 induced monokine release (TNF-α, IL-1β, IL-8, GM-CSF) in the absence of costimulus and was found to be, in the presence of monocytes, a potent costimulator of IL-2 or IL-12-induced IFNγ production by resting T cells. In contrast, in the mouse, the cytokine-like activities of sCD23 were not observed.[77]

To assess roles of CD23 in lymphocyte development and immune function in vivo, CD23-deficient and CD23 transgenic mice were generated. CD23-deficient mice display normal lymphocyte differentiation and can mount normal polyclonal IgE antibody responses.[78] One of the roles of CD23 in humans and mice is the IgE-dependent focusing Ag presentation to T cells. Thus B cells can present antigens to T cells 100x more efficiently in the presence than in the absence of IgE.[79] Antigen-specific IgE-mediated enhancement of antibody responses is severely impaired in CD23 deficient animals suggesting that IgE-increased Ag presentation by B cells may function in vivo.[80,81]

Moreover, Yu et al described that disruption of the CD23 gene led to increased and sustained specific IgE antibody titers, after immunization with thymus-dependent antigen[81] suggesting that the in vivo role of CD23 lies in negative feed-back control of IgE. The binding of IgE-containing immune complexes to CD23 may render the B cells unresponsive to IgE, as previously shown in vitro. Finally, transgenic mice with 38 kDa sCD23 over-expression in T and B cell compartments had no significant alterations in lymphoid cell maturation or in IgG1 or IgE levels in serum.[78]

OTHER SOLUBLE FCR

Soluble forms for FcαR and FcδR have been described in supernatants of cells of the immune system. These IgA-binding factors (IgA-BF)[82-85] and IgD-binding factors[86] (IgD-BF) were found to exert a regulatory roles in the isotypic control of antibody responses but have not been identified molecularly as yet.

REFERENCES

1. Fridman WH, Golstein P. Immunoglobulin-binding factor present on and produced by thymus-processed lymphocytes (T cells). Cell Immunol 1974; 11:442-455.
2. Néauport-Sautès C, Rabourdin-Combe C, Fridman WH. T cell hybrids bear Fcγ receptors and secrete suppressor immunoglobulin binding factor. Nature 1979; 277:656-659.
3. Pure E, Durie CJ, Summerill CK et al. Identification of soluble Fc receptors in mouse serum and the conditioned medium of stimulated B cells. J Exp Med 1984; 160:1836-1849.
4. Esposito-Farèse ME, Sautès C, de la Salle H et al. Membrane and soluble FcγRII/III modulate the antigen presenting capacity of murine dendritic epidermal Langerhans cells for IgG-complexed antigens. J Immunol 1995; 155:1725-1736.
5. Teillaud J-L, Bouchard C, Astier A et al. Natural and recombinant soluble low-affinity FcγR: detection, purification, and functional activities. Immuno Meth 1994; 4:48-64.
6. Daëron M, Sautès C, Bonnerot C et al. Murine type II Fcγ receptors and IgG-binding factors. Chem Immunol 1989; 47:21-78.
7. Lynch A, Tartour E, Teillaud JL et al. Increased levels of soluble low affinity FcγReceptor (IgG-binding factor) in the sera of tumour-bearing mice. Clin Exp Immunol 1992; 87: 208-214.
8. Sautès C, Varin N, Teillaud C et al. Soluble Fcγ receptors II (FcγRII) are generated by cleavage of membrane FcγRII. Eur J Immunol 1990; 21:231-234.
9. Fridman WH, Bonnerot C, Daëron M et al. Structural bases of Fcγ receptor functions. Immunol Rev 1992; 125:49-76.
10. Tartour E, de la Salle H, de la Salle C et al. Identification, in mouse macrophages and in serum, of a soluble receptor for the Fc portion of IgG (FcγR) encoded by an alternatively spliced transcript of the FcγRII gene. Intern Immunol 1993; 5:859-868.

11. Sautès C, Galinha A, Bouchard C et al. Recombinant soluble Fcγ receptors: production, purification and biological activities. J Chromatography B 1994; 662:197-207.

12. Varin N, Sautès C, Galinha A et al. Recombinant soluble receptors for the Fcγ portion inhibit antibody production in vitro. Eur J Immunol 1989; 19:2263-2268.

13. Hindley AS, Gao Y, Nash PH et al. The interaction of IgG with FcγRII: Involvement of the lower hinge binding site as probed by NMR. Bioch Soc Transactions 1993; 21:337S.

14. Lund J, Pound JD, Jones PT et al. Multiple binding sites on the CH$_2$ domain of IgG mouse FcγRII. Mol Immunol 1992; 29:53-59.

15. Diamond B, Birshtein BK, Scharff MD. Site of binding of mouse IgG2b to the Fc receptor on mouse macrophages. J Exp Med 1979; 150:721-726.

16. Sandor M, Galon J, Takacs L et al. An alternative Fcγ-receptor ligand: Potential role in T cell development. Prod Natl Acad Sci USA 1994; 91:12857-12861.

17. Sarmay G, Rozsnyay Z, Gergely J. FcγRII expression and release on resting and activated human B lymphocytes. Mol Immunol 1990; 27:1195-1200.

18. Astier A, de la Salle H, de la Salle C et al. Human epidermal Langerhans cells secrete a soluble receptor for IgG (FcγRII/CD32) that inhibits the binding of immune-complexes to FcγR+ cells. J Immunol 1994; 152:201-212.

19. Gachet C, Astier A, de la Salle H et al. Release of FcγRIIa2 by activated platelets and inhibition of anti-CD9-mediated platelet aggregation by recombinant FcγRIIa2. Blood 1995; 85:698-704.

20. Rappaport EF, Cassel DL, Walterhouse DO et al. A soluble form of the human Fc receptor FcγRIIA: cloning, transcript analysis and detection. Exp Hematol 1993; 21:689.

21. Teillaud C, Fridman WH, Sautès C. Mise en évidence de récepteurs solubles de type II pour la partie Fc des IgG (ou RFcγIIs) dans la salive totale humaine. J Biol Buccale 1992; 20:3-10.

22. Astier A, de la Salle H, Moncuit J et al. Detection and quantification of secreted soluble FcγRIIA in human sera by an enzyme-linked immuno-adsorbent assay. J Immunol Methods 1993; 166:1-10.

23. Sarmay G, Rozsnyay Z, Szabo L et al. Modulation of type II Fcγ receptor expression on activated human B lymphocytes. Eur J Immunol 1991; 21:541-549.

24. Walterhouse DO, Cassel DL, Schreiber AD et al. Characterization of HEL cell FcγRII cDNA clone lacking the sequence coding for the transmembrane region. Blood 1988; 72:344a.

25. Warmerdam PAM, Van de Winkel JGJ, Vlug A et al. A single amino acid in the second domain of the human Fcγ Receptor II is critical for human IgG2 binding. J Immunol 1991; 147:1338-1343.

26. de la Salle C, Esposito-Farèse ME, Bieber T et al. Release of soluble FcγRII/CD32 molecules by human Langerhans cells: a subtle balance between shedding and secretion ? J. Investig Dermatol 1992; 99:15s-17s.

27. de la Salle H, Astier A, Esposito-Farese ME et al. Soluble FcγRII/CD32 released from human Langerhans cells are produced, at least in part, by the secretion of a transmembrane deleted receptor. Arch Derm Res 1993; 285:109.

28. Huizinga TWJ, de Haas M, van Oers MHJ et al. The plasma concentration of soluble Fc-gamma RIII is related to production of neutrophils. British Journal of Haematology 1994; 87:459-463.

29. Selvaraj P, Rosse WF, Silber R et al. The major Fc receptor in blood has a phosphatidylinositol anchor and is deficient in paroxysmal nocturnal haemoglobinuria. Nature 1988; 333:565-567.

30. Huizinga TWJ, Kuijpers RWAM, Kleijer M et al. Maternal neutrophil FcRIII deficieny leading to neonatal isoimmune neutropenia. Blood 1990; 76:1927-1932.

31. Koene HR, de Haas M, Kleijer M et al. NA-phenotype-dependent differences in neutrophils FcγRIIIb expression cause differences in plasma levels of soluble FcγRIII. Br J Haematol 1996; 93:235-241.

32. de Haas M, Kleijer M, Minchinton RM et al. Soluble FcγRIIIa is present in plasma and is derived from natural killer cells. J Immunol 1994; 152:900-907.

33. Bazil V, Strominger JL. Metalloprotease and serine protease are involved in cleavage of CD43, CD44, and CD16 from stimulated human granulocytes. J Immunol 1994; 152:1314-1322.

34. Huizinga TW, De haas M, Kleijer M et al. Soluble Fcγ receptor III (CD16) in human plasma originates from release by neutrophils. J Clin Invest 1990; 86:416-423.

35. Tosi MF, Zakem H. Surface expression of Fcγ Receptor III (CD16) on chemoattractant-stimulated neutrophils is determined by both surface shedding and translocation from intracellular storage compartments. J Clin Invest 1992; 90:462-470.

36. Müllberg J, Durie FH, Otten-Evans C et al. A metalloprotease inhibitor blocks shedding of the IL-6 receptor and the p60 TNF receptor. J Immunol 1995; 155:5198-5205.

37. Dransfield I, Buckle AM, Savill JS et al. Neutrophil apoptosis is associated with a reduction in CD16 (FcγRIII) expression. J Immunol 1994; 153:1254-1263.

38. Homburg CHE, de Haas M, von dem Borne AEGK et al. Human neutrophils lose their surface FcγRIII and acquire annexin V binding sites during apoptosis in vitro. Blood 1995; 2:532-540.

39. Harrison D, Phillips JH, Lanier LL. Involvement of a metalloprotease in spontaneous and phorbol ester-induced release natural killer cell-associated FcγRIII (CD16-II). J Immunol 1991; 147:3459-3465.

40. Teillaud C, Galon J, Zilber M-T et al. Soluble CD16 binds peripheral blood mononuclear cells and inhibits pokeweed-mitogen (PWM)-induced responses. Blood 1993; 82:3081-3090.

41. Galon J, Gauchat JF, Mazières N et al. Soluble Fcγ Receptor type III (FcγRIII, CD16) triggers cell activation through interaction with complement receptors. J Immunol 1996; 157:1184-1192.

42. Zhou M-J, Todd III RF, van de Winkel JGJ et al. Cocapping of the leukoadhesin molecules complement receptor type 3 and lymphocyte function-associated antigen-1 with Fcγ receptor III on human neutrophils. J Immunol 1993; 150:3030-3041.

43. Poo H, Krauss J, Mayo-Bond L et al. Interaction of Fc gamma receptor type IIIB with complement receptor type 3 in fibroblast transfectants: evidence from lateral diffusion and resonance energy transfer studies. J Mol Biol 1995; 247:597-612.

44. Zhou M, Brown E. CR3 (Mac-1, alpha M beta 2, CD11b/CD18) and Fc gamma RIII cooperate in generation of a neutrophil respiratory burst: requirement for Fc gamma RIII and tyrosine phosphorylation. J Cell Biol 1994; 125:1407.

45. Gisler RH, Fridman WH. Suppression of in vitro antibody synthesis by Immunoglobulin-Binding Factor. J Exp Med 1975; 142:507-511.

46. Gisler R H, Fridman WH. Inhibition of the in vitro 19S and 7S antibody response by Immunoglobulin- Binding Factor (IBF) from alloantigen-activated T cells. Cell Immunol 1976; 23:99-107.

47. Teillaud J-L, Amigorena S, Moncuit J et al. Immunoglobulin G-binding factors (IgG-BF) inhibit IgG secretion by, as well as proliferation of, hybridoma B cells. Immunol Letters 1987; 16:139-144.

48. Lê Thi Bich T, Samarut C, Brochier J et al. Suppression of mitogen-induced peripheral B cell differentiation by soluble Fcγ receptors released from lymphocytes. Eur J Immunol 1980; 10:894-899.

49. Bouchard C, Galinha A, Tartour E et al. A transforming growth factor β-like immunosuppressive factor in immunoglobulin G-binding factor. J Exp Med 1995; 182:1717-1726.

50. Fridman WH, Teillaud J-L, Bouchard C et al. Soluble Fcγ receptors. J Leuk Biol 1993; 54:504-512.

51. Aarli A, Skeie Jensen T, Ulvestad E et al. Suppression of mitogen-induced lymphoproliferation by soluble IgG Fc receptors in retroplacental serum in normal human pregnancy. Scand. J Immunol 1993; 37:237-243.

52. Stach RM, Rowley A. A first or dominant immunization. II. Induced immunoglobulin carries transforming growth factor β and suppresses cytolytic T Cell responses to unrelated alloantigens. J Exp Med 1993; 178:841-852.

53. Naito K, Hirama M, Okumura K et al. Soluble form of the human high-affinity receptor for IgE inhibits recurrent allergic reaction in a novel mouse model of type I allergy. Eur J Immunol 1995; 25:1631-1637.

54. Ra C, Kuromitsu S, Hirose T et al. Soluble human high-affinity receptor for IgE abrogates the IgE-mediated allergic reaction. Intern Immunol 1993; 5:47-54.

55. Ishizaka K. Twenty years with IgE: from the identification of IgE to regulatory factors for the IgE response. J Immunol 1985; 135:1-10.

56. Ishizaka T, Sterk AR, Daëron M et al. Biochemical analysis of desensitization of mouse mast cells. J Immunol 1985; 135: 492-501.

57. Kuff EL, Mietz JA, Trounstine ML et al. cDNA clones encoding murine IgE-binding factors represent multiple structural variants of intracisternal A-particle genes. Proc Natl Acad Sci USA 1986; 83:6583-6587.

58. Moore KW, Jardieu P, A. MJ et al. Rodent IgE-binding factor genes are members of an endogeneous retrovirus-like gene family. J Immunol 1986; 136:4283-4290.

59. Sarfati M, Nakajima T, Frost H et al. Purification and partial biochemical characterization of IgE-binding factors secreted by a human B lymphoblastoid cell line. Immunology 1987; 60:539-545.

60. Kikutani H, Inui S, Sato R et al. Molecular structure of human lymphocyte receptor for Immunoglobulin E. Cell 1986; 47:657-665.

61. Delespesse G, Sarfati M, Hofstetter H. Human IgE-binding factors. Immunol Today 1989; 10:159-164.

62. Letellier M, Nakajima T, Pulido-Cejudo G et al. Mechanism of formation of human IgE-binding factors (soluble CD23): III. Evidence for a receptor (FcεRII)-associated proteolytic activity. J Exp Med 1990; 172:693-700.

63. Matsui M, Nunez R, Sachi Y et al. Alternative transcripts of the human isoform in the type-II cell surface receptor. FEBS Letter 1993; 335:51-56.

64. Beavil A, Edmeades R, Gould H et al. α-Helical coiled-coil stalks in the low-affinity receptor for IgE (FcεRII/CD23) and related C-type lectins. Proc Natl Acad Sci USA 1992; 89:753-760.

65. Delespesse G, Suter U, Mossalayi D et al. Expression, structure and function of the CD23 antigen. Adv Immunol 1991; 49:149-191.

66. Bourget I, Di Berardino W, Breittmayer J-P et al. CD20 monoclonal antibodies decrease interleukin-4-stimulated expression of the low-affinity receptor for IgE (FcεRII/CD23) in human B cells by increasing the extent of its cleavage. Eur J Immunol 1995; 25:1872-1876.

67. Bonnefoy JY, Defrance T, Peronne C et al. Human recombinant interleukin 4 induces normal B cells to produce soluble CD23/IgE-binding factor analogous to that spontaneously released by lymphoblastoid B cell lines. Eur J Immunol 1988; 18:117-122.

68. Hashimoto S, Koh K, Tomita Y et al. TNF-α regulates IL-4 induced FcεRII/CD23 gene expression and soluble FcεRII release by human monocytes. Int Immunol 1995; 7:705-713.

69. Pène J, Rousset F, Brière F et al. Interleukin 5 enhances interleukin 4-induced IgE production by normal human B cells. The role of soluble CD23 antigen. Eur J Immunol 1988; 18:929-935.

70. Bertho J-M, Fourcade C, Dalloul AH et al. Synergistic effect of interleukin 1 and soluble CD23 on the growth of human CD4+ bone marrow-derived T cells. Eur J Immunol 1991; 21:1073-1080.

71. Mossalayi MD, Arock M, Bertho J-M et al. Proliferation of early human myeloid precursors induced by interleukin-1 and recombinant soluble CD23. Blood 1990; 75:1924-1930.

72. Mossalayi MD, Lecron J-C, Dalloul AH et al. Soluble CD23 (FcεRII) and interleukin 1 synergistically induce early human thymocyte maturation. J Exp Med 1990; 171:959-964.

73. Liu Y-J, Cairns JA, Holder MJ et al. Recombinant 25-kDa CD23 and interleukin 1α promote the survival of germinal center B cells: evidence for bifurcation in the development of centrocytes rescued from apoptosis. Eur J Immunol 1991; 21:1107-1114.

74. Aubry J-P, Pochon S, Graber P et al. CD21 is a ligand for CD23 and regulates IgE production. Nature 1992; 358:505-507.

75. Lecoanet-Henchoz S, Gauchat J, Aubry J et al. CD23 regulates monocyte activation through a novel interaction with the adhesion molecules CD11b-CD18 and CD11c-CD18. Immunity 1995; 3:119-125.

76. Armant M, Rubio M, Delespesse G et al. Soluble CD23 directly activates monocytes to contribute to the antigen-independent stimulation of resting T cells. J Immunol 1995; 155:4868-4875.

77. Bartlett C, Conrad DH. Murine soluble FcεRII: A molecule in search of a function. Res Immunol 1992; 152:3378.

78. Stief A, Texido G, Sansig G et al. Mice deficient in CD23 reveal its modulatory role in IgE production but no role in T and B cell development. J Immunol 1994; 152:3378-3390.

79. Kehry MR, Yamashita LC. Role of the low-affinity Fcε receptor in B lymphocyte antigen presentation. Res Immunol 1990; 141:77-81.

80. Fujiwara H, Kikutani H, Suematsu S et al. The absence of IgE antibody-mediated augmentation of immune responses in CD23-deficient mice. Proc Natl Acad Sci USA 1994; 91:6835-6839.

81. Yu P, Kosco-Vilbois M, Richards M et al. Negative feedback regulation of IgE synthesis by murine CD23. Nature 1994; 369:753-755.

82. Kiyono H, Mosteller-Barnum LM, Pitts AM et al. Isotype-specific immunoregulation. IgA-binding factors produced by Fcα receptor-positive T cell hybridomas regulate IgA responses. J Exp Med 1985; 161:731-747.

83. Yodoi J, Adachi M, Teshigawara K et al. T cell hybridomas coexpressing Fc receptors (FcR) for different isotypes. II. IgA-Induced formation of suppressive IgA-Binding Factor(s) by a murine T hybridoma bearing FcγR and FcαR. J Immunol 1983; 131:303-310.

84. Yodoi J, Adachi M, Keisuke T et al. T cell hybridoma co-expressing Fc receptors for different isotypes. I. Reciprocal regulation of FcαR and FcγR expression by IgA and interferon. Immunol 1983; 4:551-559.

85. Simpson SD, Snider DP, Zettel LA et al. Soluble FcR block suppressor T cell activity at low concentration in vitro allowing isotype-specific antibody production. Cell Immunol 1996; 167:122-128.

86. Coico RF, Siskind GW, Thorbecke J. Role of IgD and T cells in the regulation of humoral immune response. Immunol. Rev 1988; 105:45-68.

87. Maliszewski CR, March CJ, Schoenborn MA et al. Expression cloning of a human Fc receptor for IgA. J Exp Med 1990; 172:1665-1672.

CHAPTER 7

FC RECEPTORS AND PATHOLOGY

Jean-Luc Teillaud

INTRODUCTION

The considerable amount of data on soluble and membrane FcR functions accumulated during the last decade has opened the way to a detailed exploration of the role of these molecules in rodent and human pathologies. Two major aspects of the role of FcR in pathology have been investigated: on the one hand, many efforts have been devoted to the study of the genetics of FcR expression, as well as the biochemical and functional characteristics of these molecules in pathology. The level of circulating soluble FcR and the presence of anti-FcR antibodies in sera of patients with various diseases have also been evaluated. On the other hand, FcR have been viewed as exquisite target molecules for immuno-intervention, based on their ability to induce cytotoxicity (ADCC), phagocytosis, endocytosis, release of cytokines and inflammatory mediators, enhancement of antigen presentation, and anergy. This chapter presents and discusses the involvement of the three FcγR (FcγRI, CD64, FcγRII, CD32, FcγRIII, CD16) and of the low-affinity FcεR (FεRII, CD23) in pathology, as well as the potential use of these receptors for immuno-intervention.

GENETIC ASPECTS OF FCR EXPRESSION: RELATIONSHIP TO DISEASES

MOUSE FCR

Autoimmunity

The genes encoding the mouse low-affinity FcγRII (*Fcgr2*) and FcγRIII (*Fcgr3*) are located on the distal part of chromosome 1,[1-3] while

Cell-Mediated Effects of Immunoglobulins, edited by Wolf Herman Fridman and Catherine Sautès. © 1997 R.G. Landes Company.

the gene encoding the high-affinity FcγRI (*Fcgr1*) is encoded on a segment of chromosome 3 between D3Nds6 (*IL-2*) and D3Nds9 (*Adh-1*).[4] Genetic studies aimed at defining a genetic linkage and tumor susceptibility genes in autoimmune diabetes and plasmacytomagenesis, respectively, have suggested that FcγR could be involved in the physiopathology of these diseases (Table 7.1).[4,5]

A congenic, non-obese diabetic (NOD) mouse strain, less susceptible to diabetes than NOD mice, was selected that contains a segment of chromosome 3 with three marker loci [D3Nds7 (*Cacy*), D3Nds11 (*Fcgr1*), and D3Nds8 (*Tshb*)]. This 7-cM region shows strong linkage to diabetes and insulitis. Interestingly, phenotypic analyses of NOD mice looking for single, fully penetrant genes permits the demonstration that NOD mice express high levels of FcγRI on monocytes, macrophages, and neutrophils after a Complete Freund's adjuvant challenge (CFA) by comparison with non-diabetic mouse strains such as C57Bl/10SnJ. This phenotype prompted Prins et al[4] to sequence the FcγRI gene of NOD and C57Bl/10SnJ mice. The B10 sequence was identical to the published one, whereas the NOD coding sequence showed several major differences. A stop codon was found at position 337, which results in a cytoplasmic tail lacking 73 amino acid residues (i.e., 75% of the total length). In addition, 24 single-base differences leading to 17 amino acid changes were also observed. This was related to a 73% reduction in the turn-over of FcγRI-IgG containing complexes present on NOD macrophages. Linkage of mutated FcγRI to insulitis and diabetes is due to increased homozygoty: one dose of the wild-type allele in heterozygous mice is sufficient to provide protection from diabetes. The role of this specific defect on NOD macrophages is not elucidated so far. One hypothesis advanced by Prins et al[4] is increased binding of NOD FcγRI to immune complexes or to monomeric IgG2a, followed by a strong ADCC and/or by the release of cytokines such as TNF-α, that is detrimental to β cells in vitro. It should be noted that amongst the 22 strains tested, only one other was found to carry the *Fcgr1* nonsense mutation, the high antibody responder Biozzi strain (AB/H).[4] A possible underlying mechanism accounting for a role of FcγRI in autoimmune diabetes could therefore be the potentiation of immune responses through the targeting of NOD macrophage FcγRI by IgG-containing immune complexes directed to β cells. Enhancement of antibody response has recently been reported when an anti-human FcγRI antibody containing known antigenic determinant was injected into human FcγRI transgenic mice.[6] However, the involvement of FcγR in autoimmune diabetes may not be restricted only to FcγRI. Recent data obtained by Luan et al in collaboration with our laboratory (manuscript submitted) indicate that a quantitative trait locus for increased IgG1 and IgG2b serum levels in the NOD mouse strain is mapped into the distal chromosome 1 region, close to the *Fcgr2* locus encoding FcγRII. The expression of FcγRIIb2 (membrane-bound)

Table 7.1. Genetic aspects of mouse FcγR expression: relationship to diseases

FcR	Locus	Mouse strain	FcγR Phenotype	Immune Phenotype
FcγRI	Fcgr 1 (chr 3)	NOD	shorter IC tail	linkage to autoimmune diabetes
"	"	Biozzi high responders	shorter IC tail	enhancement of antibody response
FcγRII	Fcgr 2 (chr 1)	NOD	weak FcγRIIb2/b3 expression	elevated IgG1 and IgG2 serum levels
FcγRII	Fcgr 2 (chr 1)	DBA/2N	wild-type	linkage to a tumor susceptibility gene (plasmacytomagenesis)

and FcγRIIb3 (soluble) isoforms of this receptor is strongly decreased on NOD macrophages, leading to a poor binding of IgG1 and IgG2b, which parallels an elevated serum level of these two subclasses. Overall, these observations indicate that the impairment of FcγR expression and functions observed could have important consequences in the regulation of immune responses in NOD mice, with possible implications in the physiopathology of autoimmune diabetes (Table 7.1). Recent studies performed on FcγRII-deficient mice showed that elevated immunoglobulin levels in response to both thymus-dependent and thymus-independent antigens are observed in these animals.[7] In addition, mast cells from these FcγRII[-/-] mice were found to be highly sensitive to IgG-triggered degranulation, confirming the down regulatory effect of FcγRII on other immune receptors.[8,9]

Plasmacytomagenesis

Fcgr2 has also been linked to a tumor susceptibility gene (Table 7.1).[5] The genetic mapping of tumor susceptibility genes involved in mouse plasmacytomagenesis helped to identify a 32-cM stretch of mouse chromosome 4 near Gt10 (gene trap insertion 10) (>95% probability of linkage) as well as a susceptibility gene on chromosome 1, likely linked to fcgr2 (90% probability of lineage). χ^2 and lod score analyses of susceptible and resistant backcross progeny between BALB/c and DBA/2N mice for 10 markers spanning chromosome 1 showed a potential linkage of this susceptibility gene to Fcgr2. It has been suggested that the DBA/2N allele of this gene is associated with susceptibility and that the BALB/c allele is associated with resistance.[5] Interestingly enough, the mouse chromosome 1 shares extensive linkage homology with stretches of the human chromosome 1 previously associated with cytogenetic abnormalities in multiple myeloma. This has led to the proposition that a tumor suppressor gene may reside on human chromosome 1 close to *Fcgr2*.[5,10] A direct involvement of low-affinity FcγR in the physiopathology of mouse plasmacytomas has not been demonstrated so far. However, two sets of data indicate that a careful examination of FcγR expression and functions should still be performed in this pathology: first, it has been shown that the expression of low-affinity FcγR on CD2[+] cells (NK and T cells) is increased in myeloma-bearing mice or in patients with IgG-producing multiple myeloma (MMγ).[11] In addition, a stage-related decrease of circulating soluble FcγRIII (sCD16), which lies as its murine counterpart close to the locus encoding FcγRII on human chromosome 1, has been reported in multiple myeloma patients (see below).[12,13]

HUMAN FcR

In man, all three FcγR are localized on chromosome 1 (FcγRI, CD64: 1q21; FcγRII, CD32: 1q23-24; FcγRIII, CD16: 1q23-24).[14,15] In addition, the human T cell receptor (TcR) ζ/η gene which encodes the ζ molecule which is also associated to NK cell FcγRIII,[16] has been linked to the FcγRII-FcγRIII gene cluster on chromosome 1q.[17] Also, both the α and γ chains of FcεRI (the γ chain being also associated to human FcγRI and FcγRIII as homo- or heterodimers with the ζ chain) have been mapped to chromosome 1q22-23.[18]

FcγRI (CD64)

Studies of familial FcγR-related defects as well as of the impact of allelic forms of FcγR on humoral and cellular immunity have provided some insights into the role of FcγR in pathology (Table 7.2). Four individuals have been identified within a single family who fail to express detectable levels of FcγRI (CD64) on their monocytes and macrophages.[19,20] Although their monocytes are unable to support mouse IgG2a anti-CD3-induced T cell proliferation, these individuals are apparently healthy. Thus, failure to express FcγRI does not provoke an entire blockade of phagocytosis in these individuals and has no significant detectable consequences on their clinical condition, especially with regard to infectious diseases. By contrast, it is interesting to point out that targeted disruption of the mouse FcγR-associated γ chain, that affects both FcγRI and FcγRIII expression on macrophages and NK cells, results in immunocompromised animals with an absence of phagocytic activity and a severe impairment of ADCC activity.[21] Thus, the presence of several FcγR with overlapping functional activities probably prevents major dysfunction of immunity in these FcγRI-defective individuals. Molecular analyses showed that all three

FcγRI genes are present, but that only one of the two mRNA species observed in normal donors (1.7 and 1.6 kb) can be detected in each of the four individuals studied (the 1.6 kb species). In addition, a single nucleotide difference within the extracellular domain exon 1-encoding region of FcγRIA has been evidenced, resulting into a termination codon (codon 92).

FcγRII (CD32)

Allelic forms of one of the FcγRII (FcγRIIa) have been described that could have important consequences for the humoral immune response against infectious pathogens (Table 7.2). These allelic variants, first identified on the basis of a functional polymorphism for the binding of mouse IgG1 and of human IgG2, have been termed the high responder (HR) or the low responder (LR) isoforms of FcγRIIa, depending on their ability to activate or not activate T cells via an IgG1 anti-CD3 mouse monoclonal antibody.[22] The HR isoform does not significantly bind human IgG2 by contrast to the LR isoform.[23,24] The molecular basis of this polymorphism has been elucidated.[24] A two-residue difference in the extracellular region has been demonstrated between HR and LR isoforms. The HR isoform contains an arginine at position 131 whereas a histidine is found in the LR isoform. This residue position is critical for the binding of both mouse IgG1 and human IgG2. A change in another position (27), tryptophan being replaced by a glutamine, has been described in both isoforms and does not play any role in this polymorphism. These observations have led to the suggestion that HR subjects could be more prone to recurrent bacterial infections due to their inability to bind human IgG2, the predominant isotype of the humoral response against bacterial capsular polysaccharides. The distribution of LR and HR isoforms, based on differential binding of murine IgG1, is 70% HR and 30% LR in Caucasians, whereas it is reversed in Asians, with 15% HR and 85% LR.[25] PCR-based sequence analysis of genomic DNA has also been used to determine the distribution in healthy individuals.[26] For Caucasian Americans, the distribution is 19% HR/HR, 51% LR/HR and 30% LR/LR, while the distribution was determined to be 26% HR/HR, 60% HR/LR and 14% LR/LR for Africans/Americans. The clinical significance of this polymorphism has been examined in different situations. Forty-eight children with recurrent bacterial respiratory tract infections were analyzed for their FcγRIIa phenotypes.[27] Interestingly enough, the LR/LR phenotype (efficient IgG2 binding) was less than half that observed in a cohort of 123 healthy adults. Conversely, in a retrospective study, 11 of 25 children who survived fulminant meningococcal septic shock were HR/HR (poor IgG2 binding) which is a significantly more frequent rate (44%) than found in healthy Caucasians (23% in this study).[28] In addition, neutrophils expressing the HR phenotype phagocytized *N. meningitidis* opsonized with polyclonal

IgG2 antibodies less effectively than did LR neutrophils in this study.[28] Thus, although a significant correlation between the failure to bind IgG2 through FcγRIIa (HR phenotype) and a higher sensitivity to recurrent bacterial infections could not be strictly demonstrated, these data suggest that FγRIIa could play an important role in host defense against bacterial infections among other factors. For instance, it has been claimed that the virtual complete absence of *Hæmophilus influenzæ* infections in Japan could be due to the predominance of the LR phenotype in Asian populations, allowing the development of IgG2-mediated immune mechanisms, the IgG2 isotype being the main IgG subclass produced following *Hæmophilus influenzæ* infection.[29]

FcγRIII (CD16)

Studies on genetic defects affecting FcγRIII (CD16) expression have shown that some individuals fail to express FcγRIIIb on neutrophils (Table 7.2). In one study,[30] 21 FcγRIIIb negative donors were genotyped for the NA polymorphism of the FcγRIIIB gene (NA-1/NA-2). All the donors were found negative for both the NA-1 and the NA-2 allele. RFLP analysis confirmed the absence of the FcγRIIIB alleles. An additional deletion of the next more telomeric located FcγRIIC gene was also found. Fourteen of the 21 patients never suffered from serious infections, while two had an autoimmune thyroiditis, four from multiple episodes of infection, and three from incidental infections. Thus, as already observed with the high-affinity FcγRI (see above), failure to express one of the two low-affinity FcγR does not provoke marked immunocompromise. It should be noted that in the latter study, the phenotype distribution of FcγRIIa was normal, excluding

Table 7.2. Genetic aspects of human FcγR expression: relationship to diseases

FcR	Locus	FcγR phenotype	Disease
FcγRI (CD64)	1q21 (chr 1)	no expression (gene deletion)	no disease observed
FcγRII (CD32)	1q23-24 (chr 1)	LR/HR allelism (HR: no IgG2 binding)	HR prone to recurrent bacterial infections?
FcγRIII (CD16)	1q23-24 (chr 1)	no expression (gene deletion)	transient neutropenia in newborn infants (allo-immunization)
"	"	NA-1/NA-2 allelism of FcγRIIIb (PMN cells)	lower IgG1/IgG3-mediated NA-2 phagocytosis?

any compensatory phenomenon based on a more frequent LR phenotype in this group of subjects.[30] Failure to express FcγRIIIb on neutrophils, also due to the absence of the FcγRIIIB gene, could be reversed in a patient in first remission of acute myeloid leukemia whose neutrophils were found to lack NA-1 and NA-2 alloantigens: bone marrow transplantation from an HLA-identical sibling allowed to convert this NA null phenotype to the normal phenotype.[31] Also, soluble FcγRIII (sCD16) was absent in the pretransplant plasma of the patient, while 20 units of sFcγRIII were detected 160 days after bone marrow transplantation. Whether this patient subsequently developed antibodies directed against FcγRIIIb has not been documented. However, in five other studies, it has been demonstrated that a moderate neutropenia observed in newborn infants was due to the presence of antibodies to FcγRIIIb, whose expression was undetectable on the mother neutrophils. These antibodies were specifically directed against FcγRIIIb. They did not bind FcγRIIIa, which was normally expressed on the maternal lymphocytes. The neutropenia was transient (reported as a transient neonatal alloimmune neutropenia, NAIN) and rapidly reverted in the children. Strikingly, the absence of FcγRIIIb on the maternal neutrophils, due to FcγRIIIB gene deletion or abnormality, was not associated with any pathology or susceptibility to infections.[32-36] The gene frequency of the NA-null phenotype was calculated as 0.0274 ± 0.0059.[34] No elevated level of circulating immune complexes was observed.[32] One patient with systemic lupus erythematosus (SLE) has also been reported to lack FcγRIIIb expression on neutrophils, but without any evidence for the implication of this defect in the pathology.[37] The absence of FcγRIIIb expression on neutrophils has also been reported in patients with paroxysmal nocturnal hemoglobinuria (PNH).[38] In that case, it is due to an acquired abnormality of hematopoietic cells affecting phosphatidylinositol glycan (PIG) membrane anchoring. By contrast to the observations reported above, this deficiency is associated with high levels of circulating immune complexes and susceptibility to bacterial infections.[38] However, the lack of FcγRIIIb expression could be unrelated to these clinical observations, as this PIG-membrane anchoring deficiency also affects other PI-linked molecules (such as decay accelerating factor or DAF, and LFA-3) that could play an important role in the clinical effects observed in PNH patients.

The role of the NA-1/NA-2 allelism[39] in the efficiency of IgG subclass-mediated phagocytosis of various bacteria and on the IgG-mediated rosette formation and phagocytosis of Rhesus D+ human red cells has also been compared.[40] The molecular basis of this polymorphism has been elucidated.[41,42] When FcγRIII is deglycosylated, the NA-1 form exhibits a mass of 29 kDa, the NA-2 form a mass of 33 kDa. The polymorphism is due to a four-amino-acid difference between the two forms, which results in the loss of two-linked glycosylation sites in the NA-1 form (four sites vs six sites). The phenotypic frequencies of the

NA-1 and NA-2 forms in Caucasians are 37% and 63%, respectively.[43] Interestingly, IgG1-mediated phagocytosis of *Staphylococcus aureus* strain Wood, *Hæmophilus influenzæ* type b, and *Neisseria meningitidis* group B has been found lower with NA-2$^{+/+}$ neutrophils than with NA-1$^{+/+}$ neutrophils. Similarly, IgG3 anti-D-mediated rosette formation and phagocytosis was also lower with NA-2$^{+/+}$ neutrophils.[40] Although this study shows an influence of the FcγRIII NA-1/NA-2 allotypes in functional interactions with biologically relevant IgG subclass antibodies (Table 7.2), whether it is related to a differential ability to mount immune responses to bacterial pathogens remains to be established.

MEMBRANE-BOUND FcγR: INVOLVEMENT IN AUTOIMMUNITY AND INFLAMMATION

MOUSE FcγR

The interaction of IgG-containing immune complexes with FcγRs expressed on various cell types results in a variety of events, from internalization of immune complexes eventually followed by the antigen processing and presentation[44] to the release of cytokines or inflammatory mediators.[15,45-54] Thus, autoantibodies directed against FcγR could trigger pathogenic events, as has been previously shown for antibodies directed against other surface receptors such as anti-acetylcholine receptor antibodies.[55] High levels of anti-FcγR autoantibodies have been found in mice prone to autoimmune diseases (NZB, NZB/NZW, Tightskin).[56] Monoclonal IgM anti-FcγR antibodies derived from these animals comprise a large subset of polyspecific IgM mAbs (60%). Both serum from these mice and anti-FcγR IgM mAbs were able to inhibit the binding of immune complexes to macrophages.[56] These antibodies could be responsible for the impairment of the macrophage functions observed in these animals.[57] In addition, these anti-FcγR antibodies induce the release of hydrolases from both azurophilic and specific granules of human neutrophils.[58] This release occurs at very low concentrations of IgM mAbs and is likely to be important in inflammation accompanying autoimmunity.

HUMAN FcγR

Similar observations have also been made in humans. For instance, a monoclonal IgG2 anti-FcγRIII antibody which has been derived from a patient with progressive systemic sclerosis, triggers the release from neutrophils of β-glucuronidase, arylsulfatase and alkaline phosphatase.[59] The presence of anti-FcγR antibodies has been reported in patients with juvenile neutropenia (see above), SLE, and localized and systemic scleroderma.[32-37,60,61] A detailed analysis of the serum from 147 patients with different systemic autoimmune diseases (SLE, Sjögren's syndrome, and progressive systemic sclerosis) showed that different patients have autoantibody directed against each of the three human FcγR.[62] Affinity

purification of seven positive sera indicated that these anti-FcγR Ig
belonged either to IgM or IgG isotypes, with a yield of 1.5-6 μg/mL
of affinity purified protein per mL of serum. Besides activating FcγR⁺
cells, anti-FcγR antibodies could block efficient clearance of immune
complexes by the mononuclear phagocyte system, leading therefore to
a higher serum concentration of immune complexes and contributing
to immune complex deposition in the kidney. In another study, high
titers of circulating IgM reacting with both FcγRII and FcγRIII were
found in SLE and rheumatoid arthritis patients. By contrast, sera from
patients with Raynaud's syndrome showed predominantly anti-FcγRIII
IgG. Patients with progressive systemic sclerosis showed both anti-FcγRII
and anti-FcγRIII IgG and IgM. Last, many patients diagnosed with
degenerative osteoarthritis also had IgG autoantibodies, directed primarily against FcγRII. No significant incidence of anti-FcγR Ig was
observed in healthy individuals in this study.[63] Studies are currently
pursued in different laboratories for examining the correlation of anti-
FcγR autoantibodies with the clinical course of various autoimmune
diseases.

SOLUBLE FcγR AND FcεR IN PATHOLOGY: MARKERS WITH A DIAGNOSIS AND/OR PROGNOSIS VALUE?

Soluble forms of FcR (sFcR) have been described in the early 70s.[64]
They were first termed immunoglobulin binding factors (IBF), as they
retain the ability to specifically bind IgG like their membrane counterparts.[65] sFcR derived from almost all the different isotype-specific
membrane receptors were then described (sFcγR, sFcαR, sFcεR, sFcδR).[66]
They are produced either by proteolytic cleavage from their membrane
counterparts or by differential splicing of their transmembrane exon.[67]

sFcγR molecules have been found in the supernatant of cells cultured in vitro or in biological fluids such as sera or saliva.[68-75] Studies
to define the biological role of sFcγR from mouse or human origin
have led to the conclusion that they could act as potent competitors
of membrane-associated FcγR by preventing IgG-FcγR interactions that
are necessary to trigger effector functions such as antibody-dependent
cell cytotoxicity (ADCC) or capture and internalization of immune
complexes.[67,75,76] In addition, it has been claimed that sFcγR, as well as
sFcεR and sFcαR, are involved in the control of antibody production
of both normal and tumor B cells such as hybridoma B cells.[65,67,75,77-81]
sFcR have been also implicated in the control of B cell proliferation.[12,82-84]
However, this has not been proven yet using recombinant molecules,
suggesting that these regulatory effects could be mediated by other
molecules having structural and/or functional relationship with sFcγR.
Notably, it has recently been shown that TGF-β, also capable of binding the Fc region of IgG, could be responsible for the suppressive
effect on B cell proliferation first attributed to sFcγR.[85] Finally, the
ability of both membrane-associated and soluble forms of FcR to bind

ligands other than Ig such as CD11b/CD18 (CR3) and CD11c/CD18 (CR4) opens the way to a larger variety of biological effects of these molecules than previously thought.[86-90]

These different observations have brought considerable interest in studying the level of circulating FcR in biological fluids from animals or patients with various pathologies as described below and summarized Tables 7.3 and 7.4.

Table 7.3 Soluble human FcγRIII and diseases

Human FcR	Disease	Serum level
FcγRIII (CD16)	MM[a]	stage-related decrease (low or absent in stage III patients)
"	MM + IVIG (in stage III patients)	increase
"	MGUS[a]	no change
"	ITP[a]+ Fcγ fragments	increase
"	HIV[+] patients[a]	drop in AIDS patients
"	B-CLL[a]	no change
"	AML[a]	decrease
"	ARDS[a]	increase (bronchoalveolar lavage fluid)
"	autoimmune-related disorders (SLE, RA, Sjögren)[a]	increase (heterogeneous)
"	ASA[a] (antisperm antibodies)	decrease (seminal plasma)

[a]MM: multiple myeloma; MGUS, monoclonal gammopathies of unknown significance; ITP, idiopathic thrombocytopenic purpura; HIV, human immunodeficiency virus; B-CLL, B chronic lymphocytic leukemia; AML, acute myelogenous leukemia; ARDS, adult respiratory distress syndrome; SLE, systemic lupus erythematosus; RA, rheumatoid arthritis; ASA, antisperm antibody syndrome.

SOLUBLE FCγR AND PATHOLOGY

Mouse soluble FcγR

In mouse, sFcγR levels are upregulated upon antigenic stimulation, allogeneic stimulation or infections with parasites (*Trypanosoma cruzi*) or viruses such as LDV (lactate dehydrogenase virus).[91,92] It is likely to reflect the stimulation of immune cells upon antigenic challenge, which leads to the release of sFcγR. This increase could be due

Table 7.4 Human soluble FcγRII and FcεRII and diseases

Human FcR	Disease	Serum level
FcγRII (CD32)	MM[a]	no change (moderate increase in stage I patients?)
"	ITP[a]+ Fcγ fragments	no change
"	B-CLL[a]	increase in stage C patients (Binet staging)
"	SLE[a]	moderate partial increase[b]
"	Behçet 's disease	moderate partial decrease[b]
FcεRII (CD23)	B-CLL	stage-related increase (elevated in stage II and III, Rai staging)
"	HIV[+] patients	increase before AIDS-associated NHL[a] appearance
"	autoimmune-related disorders (SLE, RA, Sjögren)[a]	increase
"	autoimmune-related disorders (Coeliac and Crohn's diseases)[a]	decrease

[a] MM: multiple myeloma; ITP, idiopathic thrombocytopenic purpura; HIV, human immunodeficiency virus; B-CLL, B chronic lymphocytic leukemia; SLE, systemic lupus erythematosus; RA, rheumatoid arthritis; NHL, non-Hodgkin's lymphoma.

[b] only a patient subgroup appears to exhibit a significant change in FcγRII serum levels, without any correlation with other biological or clinical parameters.

to the production of cytokines that affect FcγR expression. For instance, it has been shown that IFN-γ, a cytokine that is induced during parasitic or viral infections, upregulates both the membrane expression of FcγR[93,94] and the release of sFcγR.[95] Similarly, TNF-α increases the production of soluble FcγRIIa2 isoform which is produced by an alternative splicing of the transmembrane exon.[95] Alternatively, the increase of sFcγR levels could be related to an activation of proteases such as metalloproteases or serine proteases that have been shown to be involved in cleavage of FcγR.[96,97]

Mice bearing B cell tumors have 2-10 times more elevated sFcγR serum levels than their normal counterparts. Using an immunodot assay with the anti-mouse FcγR monoclonal antibody 2.4G2, the sFcγR levels have been measured in the sera of mice bearing syngeneic tumors of lymphoid and non-lymphoid origin.[98] These sera contain elevated amounts of sFcγR as compared to control mice. The enhancement of sFcγR levels was found to be independent on whether tumor cells express membrane-bound FcγR or not, suggesting that most of sFcγR are produced in response to tumors and originate from host rather than tumor cells. However, injection of mice with milligram amounts of purified mouse monoclonal IgG2a also provoked an increase of sFcγR serum levels, although to a lesser extent.[98] Thus, the increase of sFcγR in tumor-bearing mice has proven difficult to analyze: sFcγR can originate from both host cells and from FcγR⁺ tumor cells and its release is triggered by the presence of a monoclonal paraprotein component. In mice bearing a variant hybridoma B cell tumor producing only the κ chain, we observed that sFcγR serum level exhibits an increase similar to that observed in mice bearing the parental IgG2a κ hybridoma B cell tumor (J.L. Teillaud and J. Moncuit, unpublished observation). Another difficulty lies in the fact that only one monoclonal antibody, 2.4G2, which does not discriminate between the different mouse FcγR isoforms, is available for performing such studies.

The role of membrane-bound and soluble FcγR in tumor progression has also been studied using an experimental model of tumors developing from BALB/c 3T3 cells transformed in vitro with Polyoma virus.[99-101] It has been shown that FcγRII is expressed by a subpopulation of Polyoma virus-transformed 3T3 cells after passaging in syngeneic animals as solid tumors. This expression decreases gradually when tumor-derived cells are explanted and grown in vitro. This phenomenon is observed again when explanted cells are inoculated into mice forming second-passage tumors.[99] Interestingly, these in vivo passaged FcγRII⁺ transformed 3T3 cells exhibit a higher malignant phenotype than their FcγRII⁻ clonal ancestors or their in vivo passaged FcγRII⁻ counterparts.[100] Molecular analyses showed that it is the FcγRIIb1 isoform which is triggered upon passaging in vivo. In addition, experiments performed using Polyoma virus-transformed 3T3 cells transfected with

FcγRIIb1 cDNA showed that these cells exhibit a significantly higher tumorigenic phenotype than FcγR⁻ neo[r] transfected Polyoma virus-transformed 3T3 cells. Finally, inoculation of a mixture of FcγRIIb1⁺ and FcγRIIb1⁻ transformed 3T3 cells led to a dominance of FcγRIIb1⁺ cells into the tumor-cell population over non-FcγRIIb1 expressor cells.[101] Thus, these data suggest that FcγR could play a major role in tumor progression in vivo. A possible mechanism could be due to the release of sFcγRII from these FcγRIIb1⁺ polyoma virus-transformed 3T3 cells, creating an imbalance in the functions regulated by these molecules.[101]

Human soluble FcγR

FcR expression on peripheral blood cells and soluble FcγRIIa1/a2 (sCD32) and FcγRIII (sCD16) serum levels have also been studied in patients with multiple myeloma (MM).[11-13] It has been reported that FcR expression is increased on peripheral blood lymphocytes of these patients.[11] The type of FcR involved was related to the isotype of the secreted monoclonal paraprotein. For instance, only FcγR expression was increased in patients with MMγ, while patients with MMα exhibited higher levels of membrane-bound FcαR but not of FcγR as compared to healthy donors.[11] However, this study did not discriminate between FcγRII (CD32) and FcγRIII (CD16). A more recent study performed in our laboratory on seven MMγ patients using monoclonal antibodies that define cell subpopulations indicated the existence of a strong lymphopenia with a decrease in the absolute number of CD3⁺ cells, associated with an inverted CD4/CD8 ratio in six of these patients.[12] Similarly, a marked decrease of the absolute number of CD16⁺ and CD56⁺ cells could be observed. However, both CD16⁺ and CD56⁺ cell percentages were not increased whatever the MMγ patient tested.[12] Thus, whether some MM patients show more elevated CD16 and/or CD32 expression on their lymphocytes remains to be clearly established.

It has been possible to analyze separately sCD32 and sCD16 serum levels in MM patients (Tables 7.3 and 7.4), as monoclonal antibodies which discriminate between these human receptors are available, making it possible to develop specific quantitative assays for FcγR evaluation.[12,70,71] sCD16 are produced both by neutrophils and NK cells, although the amount of sCD16 derived from these latter cells is negligible compared with the amount derived from neutrophils.[102,103] sCD16 serum levels have been analyzed in 165 MM patients, 29 patients with monoclonal gammopathies of unknown significance (MGUS), and 20 normal disease-free donors.[13] The level of sCD16 was found to be significantly decreased in sera from MM patients compared to sera from healthy and MGUS donors (P = 0.0001). In addition, a stage-dependent decrease in sCD16 was observed, with a highly significant difference (P = 0.004) between stage I and stage II+III MM patients (Durie-Salmon staging).[13] The correlation between the myeloma stage and the serum level of sCD16, which is related to the host response (as myeloma cells express only type II FcγR, CD32) was also found more sensitive than that of β2-microglobulin, which reflects the tumor burden.[13] Interestingly, the decrease of sCD16 serum level was observed both in MMγ patients, in MMα patients, and in patients with Bence-Jones disease. By contrast to sCD16, MM patients showed no decrease of their soluble FcγRIIa1/a2 (sCD32) serum levels whatever the stage of the disease (F. Vely, unpublished observations).

The opposite modulation of FcγR serum levels observed in MM patients as compared with myeloma-bearing mice (decrease vs increase, respectively) has to be interpreted cautiously. First, both soluble FcγRII and FcγRIII serum levels were evaluated in mice with no discrimination between the two types of receptors. This quantitation reflects therefore both the tumor burden and the host immune response. By contrast, solely sCD16 (FcγRIII), which reflects the host immune response, were quantitated in MM patients. Second, the kinetics of the two diseases are different. The disease may have been first accompanied by the induction of FcγR, both in myeloma-bearing mice and in MM patients, due to the interaction with the monocloanl paraprotein, followed by a marked decrease as the disease progresses. However, the second step of the disease could be observed only in MM patients due to the rapid death of the mice. This hypothesis is strengthened by the fact that the injection in mice of monoclonal antibodies reacting both with FcγRII and FcγRIII initially induces an increase in cell-bound FcγR, followed by a decrease several days later.[92]

A decrease of sCD16 serum levels has also been reported in AIDS patients.[104] An initial increase of sCD16 serum level in clinical stage II and III (Staging of the Centers for Disease Control, CDC, 1990) is followed by a dramatic drop in patients with AIDS (stage IV). These changes correlated with the number of CD4+ cells, the amount of p24 antigen in serum, and the anti-p24 antibody titers. Thus, it has been proposed that the sCD16 serum level could be a serum marker of HIV-related disease progression.[104] No specific changes in the number of CD16+ (FcγRIIIa) natural killer cells was found, although there was a statistical correlation between the absolute number of CD3+/CD16+ cells and sCD16 serum levels.[104] Another study showed that a substantial population (25%) of neutrophils in patients with AIDS, AIDS-related complex and in HIV-1+ intravenous drug abusers does not express FcγRIIIb (CD16).[105] No changes in the expression of FcγRII (CD32), CD11b, or DAF (also a PIG-anchored molecule) were found. Thus, it was concluded that the presence of a FcγRIIIb negative neutrophil population may be related to altered functions leading to common bacterial infections in advanced AIDS.[105] However, as previously mentioned, the non-expression of FcγRIIIb due to gene deletion does not lead to recurrent bacterial infections. Thus, whether this partial deficiency of FcγRIIIb expression on neutrophils in HIV-1 patients can account for bacterial infections has still to be demonstrated.

sCD16 and sCD32 serum levels have also been investigated in a number of other diseases (Tables 7.3 and 7.4). No major changes have been observed in 40 patients with B chronic lymphocytic leukemia (B-CLL).[71] When compared to sera from healthy donors, which appear to be extremely heterogeneous in their amount of sCD16, no significant modification was noted. A number of patients had very low levels, most of this group of patients having normal lymphocyte counts. The quantification of FcγRIIa2 (the soluble form related to membrane-bound FcγRIIa1/sCD32) in a group of 57 patients with B-CLL indicated that stage C (Binet's staging) patients exhibit a significant increase of FcγRIIa2 serum levels, compared to healthy donors, stage A and B patients, or patients with complete remission (A. Astier et al, submitted). This increase may be related to the tumoral mass or to a rapid lymphocyte count doubling time, although no significant difference has been demonstrated for these parameters between C versus A and B stages. FcγRIIa2 could originate from tumor B cells (as it has been reported that FcγRIIa transcripts are present in B cell lines representing different developmental stages)[106] or from cells belonging to the host immune system. The increase of FcγRIIa2 serum level in stage C patients could reflect a worsening of their clinical conditions triggering the release of cytokines such as TNF-α which has been involved in an increased production of FcγRIIa2.[95]

A retrospective study has been performed to evaluate sCD16 serum levels in 46 patients with acute myelogenous leukemia (AML), which is characterized by granulopenia and an increase in circulating myeloblasts and occasionally promyelocytes.[107] A significantly lower concentration of sCD16 was observed in the serum of AML patients by comparison with 48 age-matched normal donors (5.4 nM vs 9.5 nM, p < 0.0005). However, it is not known whether this decrease is due to granulocytopenia. In addition, the correlation of sCD16 serum levels with the clinical status of AML patients remains to be established.[107]

Analyses of sCD16 and sFcγRIIa (sCD32) serum levels in patients with an autoimmune-related pathogeny (such as systemic lupus erythematosus, SLE, Sjögren's syndrome, rheumatoid arthritis, Behçet's disease) or with inflammatory syndromes have been also conducted (Tables 7.3 and 7.4). Heterogeneous results have been obtained, which do not define the diagnostic and/or pronostic values of these two markers. Study of a group of 50 SLE patients indicated a slight increase of sCD16 in comparison with a group of 20 normal donors.[108] However, a detailed analysis showed that the group of SLE patients is very heterogeneous, with 11 patients only having a sCD16 serum level above the chosen cutoff (0.25 µg/mL). Furthermore, four SLE patients with sicca syndrome exhibited significantly lower sCD16 serum levels than the 46 remaining SLE patients.[108] It has also been suggested that sCD16 serum levels are higher in patients with primary Sjögren's syndrome, with a concomitant decrease in the percentage of FcγRIIIb+ (CD16)

polymorphonuclear (PMN) cells. Diminished adherence and chemotaxis of PMN cells paralleled the decrease of FcγRIIIb expression.[109] The
study of 33 patients with rheumatoid arthritis (RA) indicated an increase of sCD16 serum levels, which were inversely correlated with the
mean fluorescence intensity (MFI) of FcγRIIIb⁺ PMN cells labelled with
anti-CD16 monoclonal antibodies.[110] By contrast, the density of CD11b
(CR3) and CD35 (CR1) on these PMN cells was markedly increased.
Synovial fluids from RA patients also contained higher levels of sCD16
compared to those observed in synovial fluids from non-RA patients.[110,111]
sCD32 serum levels have been examined in our laboratory in a group of
108 SLE patients and of 32 patients with Behçet's disease (P. Ghillani
and L. Musset, unpublished data). No significant difference was observed
with a group of 103 healthy donors. However, 24 of the 108 SLE
patients (22.2%) exhibited sCD32 serum levels higher than 90 ng/mL
while only 7 of the 103 normal donors (6.8%) had sCD32 values
above this level. Conversely, seven patients with Behçet's disease (21.8%)
were found with sCD32 serum levels lower than 20 ng/mL by comparison with only 14.5% of the healthy donors. No clinical or biological
correlations with known parameters could be defined in this study.

The role of sCD32 has been also examined using a recombinant
purified molecule obtained by insertion of a termination codon 5' of
sequences encoding the transmembrane domain of a human FcγRII
cDNA.[112] The administration of recombinant sCD32 significantly inhibited the immune complex-mediated inflammatory response induced
by the reversed passive Arthus reaction model in rats. The perivascular
infiltrate and the red cell extravasation was less pronounced in the
sCD32-treated group of animals. It has been also shown that recombinant sCD32 is a potent regulator of immune complex formation,
delaying immune precipitation. However, it does not inhibit the comple-
ment-mediated prevention of immune precipitation, indicating that it
does not block C1 binding to IgG containing immune complexes.[113]
Thus, sCD32 could be a valuable therapeutic agent for the treatment
of antibody or immune complex-mediated tissue damage.

sCD16 and sCD32 have also been found in other biological fluids
such as saliva, urine and seminal fluid but at much lower concentrations than that found in serum.[67,111,114,115] In patients with adult respiratory distress syndrome (ARDS), levels of sCD16 in the bronchoalveolar
lavage fluid were found five to seven times higher than that in healthy
adults.[111] A significant decrease of sCD32 levels has been reported in
saliva from patients with acute periodontitis when compared with
periodontitis-free patients, while no change in sCD16 levels was observed.[114] It has been shown that sCD16 levels are lower in seminal
plasma from patients with antisperm antibodies (ASA) than in ASA
negative patients. This may be due to a steric interference from IgG,
that could prevent sCD16 from modulating immunosuppression of
antisperm immune responses observed in these patients.[115]

Soluble FcεR and Pathology

Soluble FcεRII (sCD23)

During the last decade, several studies have indicated that soluble FcεRII (also termed IgE-Binding Factors or sCD23) may regulate IgE production by interacting primarily with IgE-bearing B cells but also with other ligands (such as CD21) expressed on a variety of cell types.[86,87,116] sCD23 can also promote the proliferation of both anti-IgM activated normal B cells and Epstein-Barr virus (EBV)-transformed B cells, acting therefore as a B cell growth factor (BCGF).[116] Thus, sCD23 serum levels have been carefully evaluated in patients with B cell-related diseases (Table 7.4).

When 40 sera from patients with B-CLL were studied, all of them showed sCD23 serum levels three to 500-fold higher than in 24 controls.[117] With a few exception, sCD23 serum levels made it possible to differentiate B-CLL from other leukemia or lymphoma patients (acute lymphocytic leukemia, multiple myeloma, non-Hodgkin's lymphoma, hairy cell leukemia). In vitro studies indicated that B lymphocytes from B-CLL patients produced eight to 50 times more sCD23 than normal B cells. sCD23 levels correlated with the Rai staging of the disease and also, although weakly, with the lymphocyte count.[117] It has been proposed that sCD23 could be not simply a B-CLL marker but may also be involved in the proliferation of the leukemic B cells through an abnormal regulation of the expression of its two membrane-bound isoforms, FcεRIIA and FcεRIIB.[118-120] Elevated sCD23 serum levels has also been described in a retrospective study of AIDS patients before the appearance of acquired immunodeficiency syndrome (AIDS) associated-non-Hodgkin's lymphoma (NHL).[121] IgE serum levels have also been found significantly elevated in these patients. Thus, serum sCD23 may serve as a clinical tool for the early detection of AIDS-associated NHL and could represent a key-component in the lymphomagenesis seen in some AIDS patients.

Other studies have shown that elevated levels of sCD23 are also found in the serum of patients with several disease states associated with elevated IgE[122,123] or with enhanced B cell activation and humoral immunity such as Sjögren's syndrome, SLE, RA, autoimmune thyroiditis, myasthenia gravis, mixed connective tissue disease and allergies (Table 7.4).[122-127] By contrast, sCD23 serum levels were found to be significantly diminished in coeliac or Crohn's disease.[125] It has been shown that increased levels of sCD23 in RA patients are related to disease status by studying monozygotic twins discordant for RA.[128] Paired analysis showed significantly elevated sCD23 levels in affected twins when compared with their unaffected co-twins or normal controls. This increase was not related to disease duration.[128] Furthermore, in EBV-related disorders after liver transplantation with immunosuppression, plasma levels of sCD23 rapidly increased when clinical symptoms were evident (fever, lymphocyte infiltration in liver biopsy) and remained high, although the EBV hepatitis improved.[127]

Soluble FcεRI

The high affinity FcεRI has been produced in soluble form by transfecting cells with a cDNA encoding the extracellular domains of the α chain of human FcεRI.[129] The purified recombinant soluble FcεRI inhibits the binding of mouse IgE to FcεRI+ cells, binds surface IgE expressed on B lymphoma cells in vitro and inhibits the passive cutaneous anaphylaxis model in vivo in Sprague-Dawley rats. Thus, these soluble molecules could represent potent regulators of type I hypersensitivity reactions. However, whether natural soluble α chain of FcεRI are present in sera of patients has not been yet documented.

MEMBRANE-BOUND AND SOLUBLE FcR: TOOLS FOR IMMUNOTHERAPY?

TARGETING FCγR WITH INTRAVENOUS IMMUNOGLOBULIN G (IVIG), FCγ FRAGMENTS, OR ANTI-FCγRIII ANTIBODY

The targeting of FcγR in diseases where autoantibodies are suspected to play a key role has been developed in the early 1980s, when it became clear that intravenous infusion of polyclonal human IgG (IVIG) in children developing acute idiopathic thrombocytopenic purpura (ITP) refractory to coticosteroid treatments allowed to obtain transient or long-lasting responses.[130] It was postulated that the blockade of FcγR present on phagocytic cells could account for these beneficial therapeutic effects, at least in part. This possibility was reinforced by the finding that IVIG slowed the clearance of radiolabelled red cells coated with anti-D in four patients with ITP and supported by the success of anti-D in raising platelet counts in D+ ITP patients.[131,132] The rationale of the treatment was that anti-D sensitized red cells would compete with IgG-coated platelets for binding to FcγR.[132] Further evidence of FcγR involvement was the efficacy of murine monoclonal antibody 3G8, directed against FcγRIII (CD16), in five of nine patients with refractory ITP and in one patient with ITP related to human immunodeficiency virus (HIV) infection.[133-135] However, the efficacy of IVIG treatment in ITP has also been related to idiotype/anti-idiotype interactions,[136] or to anti-infectious effect. These latter hypotheses have been seriously challenged, when it was shown that infusion of Fcγ fragments is also an efficient treatment of acute ITP in children.[137] Eleven of twelve children receiving Fcγ fragments (150 mg/kg daily on five consecutive days) showed a rapid increase in platelet counts to above the critical value of $50 \times 10^9/L$, thereby avoiding major hemorragic risk. Six children showed a stable response. While no sCD16 was detected in the Fcγ preparation used, sCD16 serum levels tested in five children showed transient or stable increases that correlated with the rise in platelet count.[137] Another recent assay confirmed the effectiveeess of Fcγ treatment in children with acute ITP, as the infusion of 300 mg/kg for only two consecutive days was equally efficient (Bonnet MC et al, submitted). In this trial, sCD16 serum levels were found increased in 6 of the 13 children tested. By contrast, kinetics studies of sCD32 serum levels did not show any modification during the first month following treatment (M-C Bonnet, submitted).

The mechanisms by which Fcγ fragments or IVIG can trigger rapid and, at least in about half of ITP patients, stable responses remain unclear. The rapid increase of platelet counts following IVIG or Fcγ fragment infusions is likely due to the saturation of the different FcγR, thus preventing IgG-coated platelets from being captured and phagocytosized. However, one has still to understand how these IVIG or Fcγ fragments, infused under monomeric forms, can block FcγRI, already saturated by circulating endogenous IgG, or bind to low-affinity FcγRII (CD32) or FcγRIII (CD16). Moreover, the stable responses observed are probably related to the pleiotropic effects triggered by FcγR/Fc interactions. For instance, it has been shown that the treatment of an ITP HIV⁺ patient with the anti-CD16 monoclonal antibody 3G8 not only induces a long-term correction of thrombocytopenia but also, to a lesser extent, a stabilization of CD4 lymphocytes for 18 months, a stimulation of natural killer function and an elevation of tumor necrosis factor-α (TNF-α), interferon-γ (IFN-γ) and granulocyte-macrophage colony-stimulating factor (GM-CSF).[135] Similarly, the intravenous infusion of polyclonal human IgG in seven patients with stage III refractory MMγ has led to a disappearance of bone pains, an improvement of the performance status (Karnovsky), and an increase of the mean survival. sCD16 which were undetectable before therapy appeared in the sera from the seven patients and peaked after two to four weeks of treatment.[138] Thus, IVIG or Fcγ fragment infusion triggers a cascade of events, from changes in the network of membrane-bound FcγR and sFcγR to cytokine or inflammatory mediator release, with important immunoregulatory consequences.

All these data suggest that the use of molecules other than IVIG able not only to target FcγR but also to discriminate amongst the different FcγR types and isoforms is certainly a therapeutic approach to carefully consider both in autoimmune diseases and in other disorders as described below.

INTERACTION OF ANTI-CD ANTIBODIES WITH FcγR VIA THEIR Fc REGION: IMPLICATIONS FOR THERAPY

The use of monoclonal antibodies of IgG isotype should be carefully considered as it has been shown that the Fc region of these antibodies can trigger important secondary events through interaction with FcγR expressed by targeted or surrounding cells.

Anti-CD19 antibody and B cell malignancies

FcγRII (CD32) expressed on malignant B cells influences the modulation induced by anti-CD19 monoclonal antibodies.[139] An anti-CD19 IgG1 monoclonal antibody was found to induce modulation of CD19 antigens on Daudi cells more rapidly than did its IgG2a switch variant. FcγRII did not co-modulate with CD19, but the anti-CD19 increased the capping and subsequent modulation of CD19 by increasing the calcium mobilization in these cells.[139] Thus, any therapy of patients with B cell malignancies based on the use of a monoclonal IgG antibody should carefully consider the IgG subclass to work with.

Anti-CD9 antibody and platelet aggregation

Activation of human platelets by IgG antibodies usually depends on their binding both to the target antigen and to FcγRII.[140] For instance, anti-CD9 monoclonal antibodies induce platelet aggregation only when their Fc region is bound to FcγRIIa1 (CD32), the only FcγR expressed on platelets.[141] Whether autoimmune disorders with anti-platelet IgG antibodies, leading to thrombocytopenia, involve such a mechanism remains to be elucidated. If so, specific targeting of FcγRIIa1 may be a way to prevent platelets from being activated by auto-antibodies. In addition, the activation of platelets by thrombin triggers the release of FcγRIIa2 (sCD32), containing the extracellular and intracellular regions of FcγRIIa1, but lacking the transmembrane domain, due to the alternative splicing of the transmembrane-coding exon.[142] FcγRIIa2 (sCD32) is likely to play an important role in the regulation of platelet activation by immune complexes or auto-antibodies, as its recombinant form competes efficiently with its membrane-bound counterpart, FcγRIIa1, for the Fc-dependent anti-CD9 antibody-induced platelet aggregation.[142]

Anti-CD3 antibody and immunosuppression in allograft transplantation

The use of the anti-CD3 antibody OKT3 as an immunosuppressive agent in clinical transplantation to prevent or to treat allograft rejection, has a major drawback, an associated cytokine release syndrome due to the binding of the Fc region of OKT3 to FcγR+ surrounding cells. Recent studies showed that FcγR binding is not essential for attaining the immunosuppressive property of OKT3, but that the IgG subclass of the Fc region is essential to its acute toxicity and immunogenicity.[143,144] Chimeric antibody containing the mouse IgG3 region that does not bind mouse FcγR did not trigger cytokine production, TcR desensitization or humoral response, while still retaining potent suppressive properties in mice.[143] Thus, the generation of a chimeric anti-CD3 monoclonal antibody containing the Fc region of a human IgG subclass with poor affinity for human FcγR (i.e., IgG2 and/or IgG4) may be beneficial in clinical transplantation.

Targeting FcγR for Inducing Immune Suppression and/or Anergy

Targeting of FcγR by IgG-containing immune complexes can also induce suppression of immune responses. It has been shown that passively administered anti-Rhesus D (RhD) IgG antibodies to RhD⁻ women immediately after delivery of an RhD[+] infant prevents Rh-immunization.[145] Furthermore, the sole prenatal treatment for severe Rhesus immunization is high-dose IVIG infusion.[146] The mechanisms by which the suppression of the anti-Rh immune response occurs are not elucidated and are a matter of debate. One could expect that the capture of anti-RhD sensitized red cells by phagocytic cells through FcγR would lead to increased presentation[44] and thus immunization. The opposite result is obtained in most treated women, with a major immunosuppression preventing any immunization. One possible basis for this immunosuppression is a cross-linking of various FcγR, or of FcγR with other FcR through IgG (and/or IgE or IgA present in trace amounts)-sensitized RhD[+] red cells. Recent data showed that cross-linking of FcεRI to low-affinity FcγRII inhibits IgE-induced release of mediator and cytokine.[8] Similarly, cross-linking of either BcR or TcR with FcγRII induces B cell[9,147,148] or T cell[9] inactivation, via the same tyrosine-based inhibition motif (ITIM) in the intra-cytoplasmic domain of FcγRIIb.[9] Another mechanism could be a release of sFcγR just after infusion, as reported in children with acute ITP treated with Fcγ fragments.[137] sFcγR (or IBF) have been shown, at least in vitro, to be involved in the immunosuppression of humoral responses[65] and thus could be to some extent be involved in this tolerization process.

TARGETING FcγR WITH BISPECIFIC ANTIBODIES

A monoclonal antibody should bring together molecules or cells that mediate the desired biological effect to be effective as a therapeutic agent. However, many potentially useful antibodies are not of the appropriate isotype and are thus unable to activate human complement and/or to trigger FcγR on human cells, which are involved in effector functions such as antibody-dependent cell cytotoxicity (ADCC) and/or internalization of immune complexes followed by antigen presentation. One experimental approach developed over the last decade to overcome this problem has been the use of bispecific antibodies composed of one anti-FcγR antibody linked to an anti-target antibody.[149] This approach has been explored mostly with anti-FcγRI (CD64) and anti-FcγRIII (CD16) bispecific antibodies, with the development of phase I trials to determine the maximum tolerated dose (MTD) and to get information about the optimal biologic dose.

It has been shown that the targeting of HIV-1 to FcγR on human phagocytes via a bispecific antibody anti-FcγRI/anti-gp120 reduces in vitro the infectivity of HIV-1 to cells from a human T cell lymphoma line, H9.[150] The addition of IFN-γ-activated PMN cells or monocytes (thus expressing high levels of FcγRI/CD64) to cultures of HIV-1 plus H9 cells in presence of the bispecific antibody provoked a marked reduction of p24 levels, below those at culture initiation. This in vitro study indicates therefore that IFNγ-activated phagocytes can affect the natural course of HIV-1 infection of T cells, which could be of potential clinical importance. However, the possibility that FcγR could also act as a enhancement factor of HIV infection in human cells has to be carefully examined. It has been reported that the addition of an anti-FcγRIII monoclonal antibody could inhibit HIV-1 and HIV-2 enhancement of infection of peripheral blood macrophages mediated by antibodies found in the blood of infected individuals and animals.[151]

Bispecific antibodies directed against FcγRI have also been developed for anti-tumor therapy. For instance, MDX-210, a bispecific antibody that binds simultaneously to FcγRI (outside the IgG binding site) and to HER-2/neu oncogene protein, was used in a phase Ia/Ib trial to determine the MTD.[152] Patients with advanced breast or ovarian cancer received a single intravenous infusion of MDX-210 at increasing dose levels from 0.35 to 10.0 mg/m^2. Treatment was well tolerated with malaise and low grade fevers. The maximum tolerated dose was 7.0 mg/m^2. The bispecific antibody could saturate FcγRI in a dose-dependent manner with up to 80% saturation at one hour. Plasma concentrations of TNF-α, IL-2 and G-CSF increased substantially after treatment. Monocytopenia occurred at 1-2 hours and resolved by 24 hours. One partial and one mixed tumor response were observed among 10 patients. However, decreased responses to MDX-210 were observed when some patients were challenged on days 8 and 15. Only weak monocytopenia and increased plasma cytokine levels were induced in that case. Human anti-mouse antibodies (HAMA) developed within 15 days in two patients, which could account, at least partly, for this desensitization.[152]

Similarly, a bispecific antibody against FcγRIII (CD16) and to HER-2/neu, 2B1, has been developed and tested in a phase-I trial.[153] Fifteen patients with c-erbB-2 overexpressing tumors were treated with 1 hour intravenous infusions of 2B1 on days 1, 4, 5, 6, 7, and 8 of a single course treatment. Doses were 1.0, 2.5 and 5.0 mg/m^2. Toxicities included fevers, rigors, nausea, vomiting, and leukopenia. The only dose-limiting effect was thrombocytopenia in two patients at the 5.0 mg/m^2 dose level. These patients had received extensive prior myelosuppressive chemotherapy. The initial treatment induced a strong increase of TNF-α, IL-6, IL-8, and, to a lesser extent, of GM-CSF and IFN-γ serum levels. HAMA were induced in 14 of 15 patients. The binding property of 2B1 appeared to be retained in vivo, as it bound to all neutrophils and to a proportion of monocytes and lymphocytes. Several minor clinical responses were observed (resolution of pleural effusions and ascites, of one liver metastasis; reduction of the thickness of chest wall disease).[153] Thus, this trial indicates that treatment with anti-FcγRIII/anti-target antigen bispecific antibodies has potent immunological consequences that could help to trigger anti-tumor functions via FcγRIII$^+$/CD16 cells in cancer patients.

The effectiveness of therapeutic approaches involving bispecific antibodies is likely to be improved in the near future. On the one hand, efforts to prevent the patients from developing HAMA have been made either by the construction of humanized anti-FcγRI[154] or FcγRIII (Dr Ring DB, Antibody Engineering meeting, La Jolla, Calif. USA, December 7-9, 1994) antibodies or by the screening of combinatorial human phage-displayed libraries to isolate recombinant antibody fragments (Fab or single chain Fv, scFv) (R. Kontermann et al, unpublished data). On the other hand, approaches to increase expression and/or functional activities of FcγR expressed by effector cells before or during bispecific antibody infusions have been explored. For instance, an elevated FcγRI expression is observed on neutrophils isolated from G-CSF-treated cancer patients.[155,156] ADCC assays demonstrated that these FcγRI⁺ neutrophils are potent anti-tumor effector cells, at least in vitro, and can be recruited by bispecific antibodies directed to FcγRI and to tumor antigens (HER-2/neu, disialoganglioside, G_{D2}).[155,156] Based on these observations, a phase I study has been developed to explore the toxicity of a G-CSF and anti-FcγRI/anti-HER-2/neu (MDX-210) bispecific antibody combination.[155] In this study, patients receiving G-CSF were treated with escalating doses of MDX-210. Some patients had transient fever and short periods of chills related to elevated plasma levels of IL-6 and TNF-α. A transient decrease of the absolute neutrophil count was also observed, as well as that of total white cell blood count. During G-CSF application, isolated neutrophils were highly cytotoxic in the presence of MDX-210 in vitro.[155] The concomitant infusion/injection of anti-FcγRIII/anti-target antigen bispecific antibody and IL-2 (that provokes an increase of FcγRIII/CD16 expression) should also be evaluated in the near future. Thus, the combined administration of a bispecific antibody against a particular FcγR and a given cytokine that upregulates this receptor may be an efficient therapeutic approach to trigger strong tumor lysis by effector cells in vivo. FcγRI targeting could be also invaluable for triggering humoral response against tumor cells. It has been recently reported that antigen targeting to this receptor, which is specifically expressed on myeloid cells, triggers enhanced antibody responses in transgenic human FcγRI/CD64 mice immunized with an anti-human FcγRI antibody containing antigenic determinants.[157]

TARGETING FcεR

Targeting FcεRI

Blockade of human FcεRI expressed on mast cells, basophils and epidermal cells could be an efficient approach for the therapy of immediate hypersensitivity reactions. One possible approach that has been evaluated is the use of peptides derived from the human ε chain of IgE and competing with IgE for FcεRI binding.[158] A recombinant peptide corresponding to residues 301-376 at the junction of constant regions 2 and 3 of the human IgE ε chain was found to block the in vivo passive sensitization of human skin mast cells and in vitro sensitization of human basophils with human IgE antibodies. However, approximately 11- to 13-fold higher concentration of the recombinant peptide than IgE myeloma protein was required for 50% inhibition of antigen-induced histamine release.[158] The design of other peptides competing also with IgE for binding to FcεRI has not been successful enough to develop therapeutic trials using this strategy so far.

Targeting FcεRII/CD23 in collagen-induced arthritis

As described above, increased serum levels of sFcεRII/sCD23 have been observed in patients with rheumatoid arthritis (RA). Thus, the effects of neutralizing anti-CD23 antibodies have been evaluated in a mouse model of type II collagen-induced arthritis.[159] Successful disease modulation was achieved by treatment of arthritic mice with either polyclonal or monoclonal antibodies. A dose-related amelioration of arthritis was achieved, with significantly reduced clinical scores and number of affected paws. A marked decrease in cellular infiltration as well as limited destruction of cartilage and bone were also evident in anti-CD23-treated animals.[159] Thus, CD23 appears to be involved in a mouse model of human RA and may be a good target for therapeutic human trials.

REFERENCES

1. Holmes KL, Palfree RGE, Hammerling U and Morse HC3d. Alleles of the Ly-17 alloantigen define polymorphisms of the murine IgG Fc receptor. Proc Natl Acad Sci USA 1985, 82:7706-10.
2. Hibbs ML, Hogarth PM and McKenzie IFC. The mouse Ly17 locus identifies a polymorphism of the Fc receptor. Immunogenetics 1985; 22:335-48.
3. Seldin MF, Roderick TH and Paigen B. Mouse chromosome 1. Mammal Gen 1991; 1:S1-17.
4. Prins JB, Todd JA, Rodrigues NR et al. Linkage on chromosome 3 of autoimmune diabetes and defective Fc receptor for IgG in NOD mice. Science (Wash DC) 1993; 260:695-98.
5. Mock BA, Krall MM and Dosik JK. Genetic mapping of tumor susceptibility genes involved in mouse plasmacytomagenesis. Proc Natl Acad Sci USA 1993; 90:9499-503.
6. Heijnen IA, van Vugt MJ, Fanger NA et al. Antigen targeting to myeloid-specific human FcγRI/CD64 triggers enhanced antibody responses in transgenic mice. J Clin Invest 1996; 97:331-8.
7. Takai T, Ono M, Hikida M et al. Augmented humoral and anaphylactic responses in FcγRII-deficient mice. Nature (London) 1996; 379:346-9.
8. Daëron M, Malbec O, Latour S et al. Regulation of high-affinity IgE receptor-mediated mast cell activation by murine low-affinity IgG receptors. J Clin Invest 1995; 95:577-85.
9. Daëron M, Latour S, Malbec O et al. The same tyrosine-based inhibition motif, in the intracytoplasmic domain of FcγRIIb, regulates negatively

BCR-, TCR-, and FcR-dependent cell activation. Immunity 1995; 3:635-46.

10. Seizinger BR, Klinger HP, Junien C et al. Report of the committee on chromosome and gene loss in human neoplasia. Cytogenet Cell Genet 1991; 58:1080-96.

11. Hoover RG, Hickman S, Gebel HM et al. Expansion of Fc receptor-bearing T lymphocytes in patients with Immunoglobulin G and Immunoglobulin A myeloma. J Clin Invest 1981; 67:308-15.

12. Teillaud JL, Brunati S, Elmalek M et al. Involvement of FcR⁺ T cells and of IgG-BF in the control of myeloma cells. Mol Immunol 1990; 27:1209-17.

13. Mathiot, Teillaud JL, Elmalek M et al. Correlation between soluble serum CD16 (sCD16) levels and disease stage in patients with multiple myeloma. J Clin Immunol 1993; 13:41-8.

14. Grundy HO, Peltz G, Moore KW, et al. The polymorphic Fcγ receptor II gene maps to human chromosome 1q. Immunogenetics 1989; 29:331-9.

15. Hulett MD and Hogarth PM. Molecular basis of Fc receptor function. Adv Immunol 1994; 57:1-127.

16. Lanier LL, Yu G and Phillips JH. Co-association of CD3 ζ with a receptor (CD16) for IgG Fc on human natural killer cells. Nature (London) 1989; 342:803-5.

17. Jensen JP, Hou D, Ramsburg M et al. Organization of the human T cell receptor ζ/η gene and its genetic linkage to the FcγRII-FcγRIII gene cluster. J Immunol 1992; 148:2563-71.

18. Le Coniat M, Kinet JP and Berger R. The human genes for the α and γ subunits of the mast cell receptor for IgE are located on human chromosome band 1q23. Immunogenetics 1990; 32:183-6.

19. Ceuppens JL, Baroja ML, van Vaecjk F et al. A defect in the membrane expression of high affinity 72 kD Fc receptors on phagocytic cells in four healthy subjects. J Clin Invest 1988; 82:571-8.

20. van de Winkel JG, de Wit TP, Ernst LK et al. Molecular basis for a familial defect in phagocyte expression of IgG receptor I (CD64). J Immunol 1995; 154:2896-903.

21. Takai T, Li M, Sylvestre D et al. FcR γ chain deletion results in pleiotropic effector cell defects. Cell 1994; 76: 519-29.

22. Tax WJM, Willems HW, Reekers PPM et al. Polymorphism in mitogenic effect of IgG1 monoclonal antibodies against T3 antigen on human T cells. Nature (London) 1983; 304:445-7.

23. Warmerdam PA, van de Winkel JGJ, Grosselin EJ et al. Molecular basis for a polymorphism of human FcγRII. J Exp Med 1990; 172:19-25.

24. Warmerdam PAM, van de Winkel JGJ, Vlug A et al. A single amino-acid in the second Ig-like domain of the human Fcγ receptor II is critical in human IgG2 binding. J Immunol 1991; 147:1338-43.

25. Abo T, Tilden AB, Balch CM et al. Ethnic differences in the lymphocyte proliferative response induced by a murine IgG1 antibody Leu4 to T3 molecule. J. Exp. Med 1984; 160:303-9.

26. Reilly AF, Norris CF, Surrey S et al. Genetic diversity in human Fc receptor II for IgG: FcγRIIA ligand-binding polymorphism. Clin Diagnostic Lab Immunol 1994; 1:640-4.

27. Sanders LA, van de Winkel JGJ, Rijkers GT et al. FcγRIIa (CD32) heterogeneity in patients with recurrent bacterial respiratory tract infections. J Infect Diseases 1994; 170:854-61.

28. Bredius RG, Derkx BH, Fijen CA et al. FcγRIIa (CD32) polymorphism in fulminant meningococcal septic shock in children. J Infect Diseases 1994; 170:848-53.

29. van de Winkel JGJ and Capel PJA. Human IgG Fc receptor heterogeneity: molecular aspects and clinical implications. Immunol Today 1993: 14:215-21.

30. de Haas M, Kleijer M, van Zwieten R et al. Neutrophil FcγRIIIb deficiency, nature, and clinical consequences: a study of 21 individuals from 14 families. Blood 1995; 86:2403-13.

31. Minchinton RM, de Haas M, von dem Borne AE et al. Abnormal neutrophil phenotype and neutrophil FcγRIII deficiency corrected by bone marrow transplantation. Transfusion 1995; 35:874-8.

32. Huizinga TWJ, Kuijpers RWAM, Kleijer M et al. Maternal genomic neutrophil FcγRIII deficiency leading to neonatal isoimmune neutropenia. Blood 1990; 76:1927-32.

33. Stroncek DF, Skubitz KM, Plachta LB et al. Alloimmune neonatal neutropenia due to an antibody to the neutrophil FcγRIII with maternal deficiency of CD16 antigen. Blood 1991; 77:1572-80.

34. Fromont P, Bettaieb A, Skouri H et al. Frequency of the polymorphonuclear neutrophil FcγRIII deficiency in the French population and its involvement in the development of neonatal alloimmune neutropenia. Blood 1992; 79:2131-4.

35. Cartron J, Celton JL, Gane P et al. Iso-immune neonatal neutropenia due to an anti-FcγRIII (CD16) antibody. Eur J Ped 1992; 151:438-41.

36. Puig N, de Haas M, Kleijer M et al. Isoimmune neonatal neutropenia caused by FcγRIIIb antibodies in a Spanish child. Transfusion 1995; 35:683-7.

37. Clark MR, Liu L, Clarkson SB et al. An abnormality of the gene that encodes neutrophil FcγRIII in a patient with systemic lupus erythematosus. J Clin Invest 1990; 86:341-6.

38. Selvaraj P, Rosse WF, Silber R et al. The major Fc receptor in blood has a phosphatidylinositol anchor and is deficient in paroxysmal nocturnal hæmoglobinuria. Nature (London) 1988; 333:565-7.

39. Huizinga TWJ, Kleijer M, Roos D et al. Differences between FcγRIII of human neutrophils and human K/NK lymphocytes in relation to the NA antigen system. In: Knapp W et al, eds. Leucocyte Typing. Oxford: Oxford University Press, 1989:582-85.

40. Bredius RG, Fijen CA, de Haas M et al. Role of neutrophil FcγRIIa (CD32) and FcγRIIIb (CD16) polymorphic forms in phagocytosis of human IgG1- and IgG3-opsonized bacteria and erythrocytes. Immunol 1994; 83:624-30.

41. Ory PA, Goldstein IA, Kwoh EE et al. Characterization of polymorphic forms of FcγRIII on human neutrophils. J Clin Invest 1989; 83:1676-81.

42. Ravetch JV and Perussia B. Alternative membrane forms of FcγRIII (CD16) on human NK cells and neutrophils: cell-type specific expression of two genes which differ in single nucleotide substitution. J Exp Med 1989; 170:481-91.

43. Lalezari P. Granulocyte antigen systems. In: Engelfriet CP et al, eds. Immunohæmatology. Amsterdam: Elsevier, 1984:33.

44. Amigorena S, Bonnerot C, Drake J et al. Cytoplasmic domain heterogeneity and functions of IgG Fc receptors in B lymphocytes. Science (Wash DC) 1992; 256:1808-12.

45. Debets JMH, van de Winkel JGJ, Ceuppens JL et al. Crosslinking of both FcγRI and FcγRII induces secretion of tumor necrosis factor by human monocytes, requiring high-affinity Fc-FcγR interactions. J Immunol 1990; 144:1304-10.

46. Simms HH, Gaither TA, Fries LF et al. Monokines released during short term Fcγ receptor phagocytosis up-regulate polymorphonuclear leukocytes and mono-phagocytic function. J Immunol 1991; 147:265-72.

47. Krutmann J, Kirnbauer R, Kock A et al. Crosslinking Fc receptors on monocytes triggers IL-6 production. J Immunol 1990; 145:1337-42.

48. Anegon I, Cuturi MC, Trinchieri G et al. Interaction of Fc receptor (CD16) ligands induces transcription of interleukin 2 receptor (CD25) and lymphokine genes and expression of their products in human natural killer cells. J Exp Med 1988; 167:452-72.

49. Anderson CL, Guyre PM, Whitin JC et al. Monoclonal antibodies to Fc receptors for IgG on human mononuclear phagocytes: antibody characterization and induction of superoxyde production in a monocyte cell. J Biol Chem 1986; 261:12856-64.

50. Willis HE, Browder B, Feister AJ et al. Monoclonal antibody to human IgG Fc receptors: cross-linking of receptors induces lysosomal enzyme release and superoxyde generation by neutrophils. J Immunol 1988; 140:234-9.

51. Tosi MF and Berger M. Functionnal differences between the 40 kDa and 50 to 70 kDa IgG Fc receptors on human neutrophils revealed by elastase treatment and antireceptors antibodies. J Immunol 1988; 141:2097-2103.

52. Daëron M, Bonnerot C, Latour S et al. Murine recombinant FcγRIII, but not FcγRII, trigger sertonin release in rat basophilic leukemia cells. J Immunol 1992; 149:1365-73.

53. Trezzini C, Jungi TW, Spycher MO et al. Human monocyte CD36 and CD16 are signalling molecules. Immunol 1990; 71:29-37.

54. Edberg JC and Kimberly RP. Modulation of Fcγ and complement receptor function by the glycosylphosphatidylinositol-anchored form of FcγRIII. J Immunol 1994; 152:5826-35.

55. Almon RR, Andrew CG and Appel SH. Serum globulin in myasthenia gravis: inhibition of alpha-bungarotoxin binding to acetylcholine receptors. Science (Wash DC) 1974; 186:55-7.

56. Boros P, Chen J, Bona C et al. Autoimmune mice make anti-Fcγ receptor Ig. J Exp Med 1990; 171:1581-95.

57. Russell PJ and Steinberg AD. Studies of peritoneal macrophage function in mice with systemic lupus erythematosus: depressed phagocytosis of opsonised sheep erythrocytes in vitro. Clin Immunol Immunopathol 1983; 27:387-402.

58. Boros P, Odin JA, Muryoi T et al. IgM anti-FcγR autoantibodies trigger neutrophil degranulation. J Exp Med 1991; 173:1473-82.

59. Szegedi A, Boros P, Chen J et al. An FcγRIII (CD16)-specific autoantibody from a patient with progressive systemic sclerosis. Immunol Lett 1993; 35:69-76.

60. Sipos A, Csortos C, Sipka S et al. The antigen/receptor specificity of antigranulocyte antibodies in patients with SLE. Immunol Lett 1988; 19:329-34.

61. Davis K, Boros P, Keltz M et al. Circulating FcγR-specific autoantibodies in localized and systemic scleroderma. J Am Acad Dermatol 1995; 33:612-6.

62. Boros P, Muryoi T, Spiera H et al. Autoantibodies directed against different classes of FcγR are found in sera of autoimmune patients. J Immunol 1993; 150:2018-24.

63. Boros P, Odin JA, Chen J et al. Specificity and class distribution of FcγR-specific autoantibodies in patients with autoimmune disease. J Immunol 1994; 152:302-6.

64. Fridman WH and Golstein P. Immunoglobulin-Binding Factor present on and produced by thymus-processed lymphocytes (T cells). Cell Immunol 1974; 11:442-5.

65. Fridman WH, Rabourdin-Combe C, Neauport-Sautès C et al. Characterization and function of T cell Fcγ receptor. Immunol Rev 1981; 56:51-88.

66. Fridman WH. Fc receptors and immunoglobulin binding factors. FASEB J 1991; 5:2684-90.

67. Teillaud JL, Bouchard C, Astier A et al. Natural and recombinant soluble low-affinity FcγR: detection, purification, and functional activities. Immunometh 1994; 4:48-64.

68. Khayat D, Dux Z, Anavi R et al. Circulating cell free Fcγ2b/γ1 receptor in normal mouse serum: its detection and specificity. J Immunol 1984; 132:2496-501.

69. Pure E, Durie CJ, Summerill CK et al. Identification of soluble Fc receptors in mouse serum and the conditioned medium of stimulated B cells. J Exp Med 1984; 160:1836-49.

70. Khayat D, Geffrier C, Yoon S et al. Soluble circulating Fcγ receptors in human serum: a new ELISA assay for specific and quantitative detection. J Immunol Meth 1987; 100:235-41.

71. Fridman WH, Mathiot C, Moncuit J et al. Fc receptors, immunoglobulin-binding factors and B chronic lymphocytic leukemia. Nouv Rev Franc Hematol 1988; 30:311-5.

72. Huizinga TWM, van der Schoot CE, Jost C et al. The PI-linked receptor FcRIII is released on stimulation of neutrophils. Nature (London) 1988; 333:667-9.

73. Lanier LL, Phillips JH and Testi R. Membrane anchoring and spontaneous release of CD16 (FcRIII) by natural killer cells and granulocytes. Eur J Immunol 1989; 19:775-8.

74. de Haas M, Kleijer M, Minchinton RM et al. Soluble FcγRIIIa is present in plasma and is derived from natural killer cells. J Immunol 1994; 152:900-7.

75. Sautès C, Teillaud C, Mazières N et al. Soluble FcγR (sFcγR): detection in biological fluids and production of a murine recombinant sFcγR biologically active in vitro and in vivo. Immunobiol 1992; 185:207-221.

76. Esposito-Farese ME, Sautès C, de la Salle H et al. Membrane and soluble FcγRII/III modulate the antigen-presenting capacity of murine dendritic epidermal Langerhans cells for IgG-complexed antigens. J Immunol 1995; 154:1725-36.

77. Gisler RH and Fridman WH. Suppression of in vitro antibody synthesis by immunoglobulin-binding factor. J Exp Med 1975; 142:507-11.

78. Varin N, Sautès C, Galinha A et al. Recombinant soluble receptors for the Fcγ portion inhibit antibody production in vitro. Eur J Immunol 1989; 19:2263-8.

79. Suemura M, Ishizaka A, Kobatake S et al. Inhibition of IgE production in hybridomas by the IgE class specific suppressor factor from T hybridomas. J Immunol 1983; 130:1056-60.

80. Simpson SD, Snider DP, Zettel LA et al. Soluble FcR block suppressor T cell activity at low concentration in vitro allowing isotype-specific antibody production. Cell Immunol 1996; 167:122-8.

81. Brunati S, Moncuit J, Fridman WH et al. Regulation of IgG production by suppressor FcγRII+ T hybridomas. Eur J Immunol 1990; 20:55-61.

82. Müller S and Hoover RG. T cells with Fc receptors in myeloma ; suppression of growth and secretion of MOPC315 by Tα cells. J Immunol 1985; 134:644-7.

83. Teillaud C, Galon J, Zilber MTh et al. Soluble CD16 binds peripheral blood mononuclear cells and inhibits pokeweed-mitogen-induced responses. Blood 1993; 82:3081-90.

84. Gordon J, Flores-Romo L, Cairns JA et al. CD23: a multi-functional receptor lymphokine ? Immunol Today 1989; 10:153-7.

85. Bouchard C, Galinha A, Tartour E et al. A Transforming Growth Factor β-like immmunosuppressive factor in Immunoglobulin G-Binding Factor. J Exp Med 1995; 182:1717-26.

86. Aubry JP, Pochon S, Graber P et al. CD21 is a ligand for CD23 and regulates IgE production. Nature (London) 1992; 358:505-7.

87. Armant M, Rubio M, Delespesse G et al. Soluble CD23 directly activates monocytes to contribute to the antigen-independent stimulation of resting T cells. J Immunol 1995; 155:4868-75.

88. Zhou MJ, Todd III RF, van de Winkel JGJ et al. Cocapping of the leukoadhesin molecules CR3 and LFA-1 with FcγRIII on human neutrophils. Possible role of lectin-like interactions. J Immunol 1993; 150:3030-41.

89. Galon J, Bouchard C, Fridman WH et al. Ligands and biological activities of soluble Fcγ receptors. Immunol Lett 1995; 44:175-81.

90. Galon J, Gauchat J-F, Mazières N et al. Soluble Fcγ receptor type III (FcγRIII, CD16) triggers cell activation through interaction with complement receptors. J Immunol 1996; 157:1184-1192.

91. Daëron M, Sautès C, Bonnerot C et al. Murine type II Fcγ receptors and IgG binding factors. Chem Immunol 1989; 47:21-78.

92. Araujo Jorge T, El Bouhdidi A, Rivera MT et al. *Trypanosoma cruzi* infection in mice enhances the membrane expression of low-affinity Fc receptors for IgG and the release of their soluble forms. Paras Immunol 1993; 15:539-46.

93. Fridman WH, Gresser I, Bandu M-T et al. Interferon enhances the expression of Fcγ receptors. J Immunol 1980; 124:2436-41.

94. Guyre PM, Morganelli PM and Miller R. Recombinant immune interferon increases immunoglobulin G Fc receptors on cultured human mononuclear phagocytes. J Clin Invest 1983; 72:393-7.

95. de la Salle C, Esposito-Farese ME, Bieber Th et al. Release of soluble FcγRII/CD32 molecules by human Langerhans cells: a subtle balance between shedding and secretion? J Invest Dermatol 1992; 99:15S-17S.

96. Harrison D, Phillips JH and Lanier LL. Involvement of a metalloprotease in spontaneous and phorbol ester-induced release of natural killer cell-associated FcγRIII (CD16). J Immunol 1991; 147:3459-65.

97. Bazil V and Strominger JL. Metalloprotease and serine protease are involved in cleavage of CD43, CD44, and CD16 from stimulated human granulocytes. Induction of cleavage of L-selectin via CD16. J Immunol 1994; 152:1314-22.

98. Lynch A, Tartour E, Teillaud JL et al. Increased levels of soluble low-affinity Fcγ receptors (IgG-Binding Factors) in the sera of tumor-bearing mice. Clin Exp Immunol 1992; 87:208-14.

99. Ran M, Katz B, Kimchi N et al. The in vivo acquisition of FcγRII expression on polyoma virus transformed cells derived from tumors of long latency. Cancer Res 1991; 51:612-8.

100. Ben-Baruch Langer A, Emmanuel A, Even J et al. Phenotypic properties of 3T3 cells transformed in vitro with polyoma virus and passaged once in syngeneic animals. J Immunobiol 1992; 185:281-91.

101. Zusman T, Gohar O, Eliassi I et al. The murine Fc-gamma (Fcγ) receptor type II B1 is a tumorigenicity-enhancing factor in polyoma-virus-transformed 3T3 cells. Int J Cancer 1996; 65:221-9.

102. Huizinga TW, de Haas M, van Oers MH et al. The plasma concentration of soluble FcγRIII is related to production of neutrophils. Brit J Hæmatol 1994; 87:459-63.

103. de Haas M, Kleijer M, Minchinton RM et al. Soluble FcγRIIIa is present in plasma and is derived from natural killer cells. J Immunol 1994; 152:900-7.

104. Khayat D, Soubrane C, Andrieu JM et al. Changes of soluble CD16 levels in serum of HIV-infected patients: correlation with clinical and biologic prognostic factors. J Inf Dis 1990; 161:430-5.

105. Boros P, Gardos E, Bekesi GJ et al. Change in expression of FcγRIII (CD16) on neutrophils from human immunodeficiency virus-infected individuals. Clin Immunol Immunopath 1990; 54:281-9.

106. van der Herik-Oudijk IE, Westerdaal NAC, Henriquez NV et al. Functional analysis of human FcγRII (CD32) isoforms expressed in B lymphocytes. J Immunol 1994; 152:574-85.

107. Fleit HB and Kobasiuk CD. Soluble FcγRIII is present in lower concentrations in the serum of patients with acute myelogenous leukemia (AML): a retrospective study. Leukemia 1993; 7:1250-2.

108. Hutin P, Lamour A, Pennec YL et al. Cell-free FcγRIII in sera from patients with systemic lupus erythematosus: correlation with clinical and biological features. Int Arch Allerg Immunol 1994; 103:23-7.

109. Lamour A, Soubrane C, Ichen M et al. FcγRIII shedding by polymorphonuclear cells in primary Sjögren's syndrome. Eur J Clin Invest 1993; 23:97-101.

110. Lamour A, Baron D, Soubrane C et al. Anti-FcγRIII autoantibody is associated with soluble receptor in rheumatoid arthritis serum and synovial fluid. J Autoimmun 1995; 8:249-65.

111. Fleit HB, Kobasiuk CD, Daly C et al. A soluble form of FcγRIII is present in human serum and other body fluids and is elevated at sites of inflammation. Blood 1992; 79:2721-8.

112. Ierino FL, Powell MS, McKenzie IF et al. Recombinant soluble human FcγRII: production, characterization, and inhibition of the Arthus reaction. J Exp Med 1993; 178:1617-28.

113. Gavin AL, Wines BD, Powell MS et al. Recombinant soluble FcγRII inhibits immune complex precipitation. Clin Exp Immunol 1995; 102:620-5.

114. Teillaud C, Jourde M, Sautès C et al. Acute periodontitis and soluble CD32 and CD16 levels in saliva. J Parodontol 1993; 12:9-18.

115. Sedor J, Callahan HJ, Perussia B et al. Soluble FcγRIII (CD16) and IgG levels in seminal plasma of men with immunological infertility. J Androl 1993; 14:187-93.

116. Delespesse G, Sarfati M and Hofstetter H. Human IgE-binding factors. Immunol Today 1989; 10:159-64.

117. Sarfati M, Bron D, Lagneaux L et al. Elevation of IgE binding factors in serum of patients with B cell-derived chronic lymphocytic leukemia. Blood 1988; 71:94-8.

118. Sarfati M. CD23 and chronic lymphocytic leukemia. Blood Cells 1993; 19:591-6.

119. Fournier S, Delespesse G, Rubio M et al. CD23 antigen regulation and signaling in chronic lymphocytic leukemia. J Clin Invest 1992; 89:1312-21.

120. Fournier S, Yang LP, Delespesse G et al. The two CD23 isoforms display differential regulation in chronic lymphocytic leukæmia. Br J Hæmatol 1995; 89:373-9.

121. Yawetz S, Cumberland WG, van der Meyden M et al. Elevated serum levels of soluble CD23 (sCD23) precede the appearance of acquired immunodeficiency syndrome-associated Non-Hodgkin's Lymphoma. Blood 1995; 85:1843-9.

122. Kim KM, Nanbu M, Iwai Y et al. Soluble low affinity Fc receptor for IgE in the serum of allergic and nonallergic children. Pediatr Res 1989; 26:49-53.

123. Yanagihara Y, Sarfati M, Marsh D et al. Serum levels of IgE-binding factor (soluble CD23) in diseases associated with elevated IgE. Clin Exp Allergy 1990; 20:395-401.

124. Bansal A, Roberts T, Hay EM et al. Soluble CD23 levels are elevated in the serum of patients with primary Sjögren's syndrome and systemic lupus erythematosus. Clin Exp Immunol 1992; 89:452-5.

125. Bansal AS, Ollier W, Marsh MN et al. Variations in serum sCD23 in conditions with either enhanced humoral or cell-mediated immunity. Immunology 1993; 79: 285-9.

126. Chomarat P, Briolay J, Banchereau J et al. Increased production of soluble CD23 in rheumatoid arthritis, and its regulation by interleukin-4. Arthritis and Rheumat 1993; 36:234-42.

127. Yoshikawa T, Nanba T, Kato H et al. Soluble FcεRII/CD23 in patients with autoimmune diseases and Epstein-Barr virus-related disorders: analysis by ELISA for soluble FcεRII/CD23. Immunomethods 1994; 4:65-71.

128. Bansal AS, MacGregor AJ, Pumphrey RS et al. Increased levels of sCD23 in rheumatoid arthritis are related to disease status. Clin Exp Rheumatol 1994; 12:281-5.

129. Gavin AL, Snider J, Hulett MD et al. Expression of recombinant soluble FcεRI: function and tissue distribution studies. Immunology 1995; 86:392-8.

130. Imbach P, Barandun S, d'Apuzzo V et al. High-dose intravenous gammaglobulin for idiopathic thrombocytopenic purpura in chilhood. Lancet 1981; i:1228-31.

131. Ferh J, Hofmann V, Kappeler U. Transient reversal of thrombocytopenia in idiopathic thrombocytopenic purpura by high dose intravenous gammma globulin. N Engl J Med 1982; 306:1254-8.

132. Salama A, Mueller-Eckhardt C, Fiefel V. Effect of intravenous immunoglobulin in immune thrombocytopenia; competitive inhibition of reticuloendothelial system function by sequestration of autologous red blood cells. Lancet 1983; ii:193-5.

133. Clarkson S, Bussel J, Kimberly R et al. Treatment of refractory immune thrombocytopenic purpura with an anti-Fcγ receptor antibody. N Engl J Med 1986; 314:1236-9.

134. Bussel J. Modulation of Fc receptor clearance and antiplatelet antibodies as a consequence of intravenous immune globulin infusion in patients with immune thrombocytopenic purpura. J Allergy Clin Immunol 1989; 84:566-78.

135. Soubrane C, Tourani JM, Andrieu JM et al. Biologic response to anti-CD16 monoclonal antibody therapy in a human immunodeficiency virus-related immune thrombocytopenic purpura patient. Blood 1993; 81:15-9.

136. Kaveri SV, Dietrich G, Hurez V et al. Intravenous immunoglobulins (IVIg) in the treatment of autoimmune disease. Clin Exp Immunol 1991; 86:192-8.

137. Debré M, Bonnet M-C, Fridman WH et al. Infusion of Fcγ fragments for treatment of children with acute immune thrombocytopenic purpura. Lancet 1993; 342:945-9.

138. Laporte J-Ph, Mathiot C, Teillaud J-J et al. Intravenous injection of polyclonal human IgG as treatment of refractory Multiple Myeloma. Blood 1990; 76 Suppl I:359a.

139. Vervoordeldonk SF, Merle PA, van Leeuwen EF et al. FcγRII (CD32) on malignant B cells influences modulation induced by anti-CD19 monoclonal antibody. Blood 1994; 83:1632-9.

140. Slupsky JR, Cawley JC, Griffith LS et al. Role of FcγRII in platelet activation by monoclonal antibodies. J Immunol 1992; 148:3189-94.

141. Worthington RE, Carroll RC, Boucheix C. Platelet activation by CD9 monoclonal antibodies is mediated by the FcγII receptor. Br J Hæmatol 1990; 74:216-22.

142. Gachet C, Astier A, de la Salle H et al. Release of FcγRIIa2 by activated platelets and inhibition of anti-CD9-mediated platelet aggregation by recombinant FcγRIIa2. Blood 1995; 85:698-704.

143. Alegre ML, Tso JY, Sattar H et al. An anti-murine CD3 monoclonal antibody with a low affinity for Fcγ receptors suppresses transplantation responses while minimizing acute toxicity and immunogenicity. J Immunol 1995; 155:1544-55.

144. Vossen AC, Tibbe GJ, Kroos MJ et al. Fc receptor binding of anti-CD3 monoclonal antibodies is not essential for immunosuppression, but triggers cytokine-related side effects. Eur J Immunol 1995; 25:1492-6.

145. Contreras M and de Silva M. The prevention and management of hæmolytic disease of the newborn. J Roy Soc Med 1994; 87:256-8.

146. de la Camara C, Arrieta R, Gonzalez A et al. High-dose intravenous immunoglobulin as the sole prenatal treatment for severe Rh immunization. N Engl J Med 1988; 318:519-20.

147. Phillips NE and Parker DC. Fc-dependent inhibition of mouse B cell activation by whole anti-μ antibodies. J Immunol 1983;130:602-6.

148. Phillipps NE and Parker DC. Cross-linking of B lymphocyte Fcγ receptors and membrane immunoglobulin inhibits anti-immunoglobulin-induced blastogenesis. J Immunol 1984; 132:627-32.

149. Fanger MW, Graziano RF and Guyre PM. Production and use of anti-FcR bispecific antibodies. Immunomethods 1994; 4:72-81.

150. Howell AL, Guyre PM, You K et al.Targeting HIV-1 to FcγR on human phogocytes via bispecific antibodies reduces infectivity of HIV-1 to T cells. J Leuk Biol 1994; 55:385-91.

151. Homsy J, Meyer M, Tateno M et al. The Fc and not CD4 receptor mediates antibody enhancement of HIV infection in human cells. Science (Wash DC) 1989; 244:1357-60.

152. Valone FH, Kaufman PA, Guyre PM et al. Phase Ia/Ib trial of bispecific antibody MDX-210 in patients with advanced breast or ovarian cancer that overexpresses the proto-oncogene Her-2/neu. J Clin Oncol 1995; 13:2281-92.

153. Weiner LM, Clark JI, Davey M et al. Phase I trial of 2B1, a bispecific monoclonal antibody targeting c-erb-2 and FcγRIII. Cancer Res 1995; 55:4586-93.

154. Graziano RF, Tempest PR, White P et al. Construction and characterization of a humanized anti-FcγRI monoclonal antibody. J Immunol 1995; 155:4996-5002.

155. Repp R, Valerius T, Wieland G et al. G-CSF-stimulated PMN in immunotherapy of breast cancer with a bispecific antibody to FcγRI and to Her-2/neu. J Hematother 1995; 4:415-21.

156. Michon J, Moutel S, Barbet J et al. In vitro killing of neuroblastoma cells by neutrophils derived from granulocyte colony-stimulating factor-treated cancer patients using an anti-disialoganglioside/anti-FcγRI bispecific antibody. Blood 1995; 86:1124-30.

157. Heijnen IA, van Vugt MJ, Fanger M et al. Antigen targeting to myeloid-specific human FcγRI/CD64 triggers enhanced antibody responses in transgenic mice. J Clin Invest 1996; 97:331-8.

158. Helm B, Kebo D, Vercelli D et al. Blocking of passive sensitization of human mast cells and basophil granulocytes with IgE antibodies by a recombinant human ε chain fragment of 76 amino acids. Proc Natl Acad Sci (USA) 1989; 86:9465-9.

159. Plater-Zyberk C and Bonnefoy J-Y. Marked amelioration of established collagen-induced arthritis by treatment with antibodies to CD23 in vivo. Nature Med (London) 1995; 1:781-5.

INDEX

DATE DUE

NOV 0 4 2001	